# Building Pathology and Rehabilitation

Volume 4

*Series editors*

Vasco Peixoto de Freitas
Aníbal Costa
J. M. P. Q. Delgado

For further volumes:
http://www.springer.com/series/10019

J. M. P. Q. Delgado
Editor

# Drying and Wetting of Building Materials and Components

Springer

*Editor*
J. M. P. Q. Delgado
LFC—Building Physics Laboratory
University of Porto
Porto
Portugal

ISSN 2194-9832 ISSN 2194-9840 (electronic)
ISBN 978-3-319-04530-6 ISBN 978-3-319-04531-3 (eBook)
DOI 10.1007/978-3-319-04531-3
Springer Cham Heidelberg New York Dordrecht London

Library of Congress Control Number: 2014942543



Printed on acid-free paper

Springer is part of Springer Science+Business Media (www.springer.com)

# Contents

# Chapter 1
# Air Drying Technologies Applied to Buildings Treatment

**J. M. P. Q. Delgado, A. S. Guimarães and V. P. de Freitas**

**Abstract** The Laboratory of Building Physics (LFC) has been developed an important research about air drying technologies applied to buildings treatment. In the last years it was validated and experimentally analysed the operating principle of a technique called "Wall Base Ventilation System" initially developed for the treatment of rising damp problems in historical and ancient building walls, consisting in use the air circulation in the base of the walls of buildings with high thickness and heterogeneity in its constitution, with a relative humidity far off saturation. The first experimental and numerical studies established allowed to value the viability of the proposed system, so it was also developed a designing methodology. Rising Damp coming from the ground by capillarity, rises through porous materials, is one of the main degradation causes of historical and ancient buildings, essentially, of its thick walls with heterogeneous composition where the traditionally techniques era sometimes ineffective or too expensive and that was the reason why it was decided to developed a new way to control it. In this work it is proposed to present two completely different applications of the "Wall Base Ventilation System" technique developed in the LFC: the relative humidity control in crawl spaces where, for a Portuguese building, it is presented the problem, the prediction of mould growth, moisture balance in the crawl space, some numerical simulations and the proposed intervention methodology and for the rising damp in walls where it is presented the problem, an experimental study and the system implementation and optimization.

J. M. P. Q. Delgado (✉) · A. S. Guimarães · V. P. de Freitas
LFC—Building Physics Laboratory, Civil Engineering Department, Faculty of Engineering, University of Porto, Porto, Portugal
e-mail: jdelgado@fe.up.pt

A. S. Guimarães
e-mail: anasofia@fe.up.pt

V. P. de Freitas
e-mail: vpfreita@fe.up.pt

J. M. P. Q. Delgado (ed.), *Drying and Wetting of Building Materials and Components*, Building Pathology and Rehabilitation 4, DOI: 10.1007/978-3-319-04531-3_1,

**Keywords** Air drying • Crawl spaces • Rising damp • Numerical simulation • Hygro-regulated system • Ventilation • Mould growth

## 1.1 Introduction

Humidity is a major cause of pathologies in old and historical building. The conservation of historical buildings has become important nowadays and has developed significantly in recent years.

Rising damp in historical buildings may be considered one of the most important of all the different manifestations of damp, leading to the decay of stone materials due to frost/defrost cycles and the presence of salts associated with crystallization, decay in elements in contact with the stone walls, such as wooden beams, wooden ceilings and other finishing's, and insanitary conditions caused by excess humidity associated with the development of fungus and mould.

The treatment of rising damp in historical buildings walls is very complex, due to the thickness of the walls and the fact that they are built from different materials. The traditional techniques used to deal with this kind of problem (such as watertight barriers, injection of hydrofuge products, etc.) sometimes prove ineffective or too expensive, justifying the need to find new approaches (see Guimarães et al. 2010).

The Building Physics Laboratory at the University of Porto-Faculty of Engineering has been developing important experimental research into the problems of rising damp. In recent years the principles for a treatment technique called the "Wall Base Ventilation System to Treat Rising Damp" has been validated and characterised in experiments. The technique consists of circulating air around the base of very thick walls built from different materials, with a relative humidity saturation distance. Wall base ventilation increases evaporation, which reduces the level of the damp front. This is possible only when the groundwater is lower than the base of the wall (Colombert 1975).

In the investigation course it was possible to understand the utility of this system for another purpose. We saw that the most important factors causing mould growth in a crawl space are relative humidity, temperature, nutrition and pH. In practice, relative humidity is the most significant factor causing mould growth in the crawl spaces. The limit value for relative humidity in crawl spaces is usually considered to be from 75 to 80 % (Samuelsson 1994). Relative humidity and temperature are strongly linked together. The thermal mass of the structures and ground soil affects the temperature behaviour of the crawl space. Reducing the moisture level in the air (relative humidity) and maintaining the inside relative humidity below 75 % it is the key to control mould growth. So, the intervention methodology proposed could be divided in two steps: control the groundwater level and ventilating the crawl spaces with the hygro-regulated system (Delgado et al. 2012; Freitas et al. 2011; Guimarães et al. 2010).

## 1.2 Air Drying Technologies

Air drying technologies to accelerate the drying process in damp spaces, walls floors, roofs or other building parts are used for a long time.

If fact, the climatic conditions of the ambiences affects the drying process and have, for example, a great influence on the level achieved by rising damp. In places with high humidity, evaporation will be with greater difficulty and consequently there will be a greater progression of damp. Otherwise, when the relative humidity is low, evaporation will be extreme and the time to progression of moisture will be lower.

In a wall, the drying flow is proportional to the concentration difference between the damp surface water vapour and the air water vapour. Inside a building, if there is no big difference between the air temperature and the inside surface of the walls, to a high relative humidity, the difference of concentration tends to zero as well as the drying flow.

Having these aspects in mind some air drying techniques were developed to control rising damp in wall building:

Reducing the Absorbent Section: This technique consists basically of diminishing the absorbent area by replacing part of the wall by air, thereby not only reducing the amount of water absorbed, but also increasing evaporation.

An Atmospheric Drainage: The principle behind atmospheric drainage is the fact that damp air is heavier than dry air. Knappen believed that inserting oblique drainage tubes in walls would release damp air (coming from inside the wall), thereby facilitating the wall-drying process.

Concealing Anomalies: When the causes of rising damp cannot be eliminated, we can decide to put up a new wall separated from the original wall with a ventilation space, which is a type of damp-proof course system concealing the anomalies. The ventilated space replaces damp air for dry air.

Knowing the limitation of the cited techniques to treat walls of old and historical building with high thickness and made by different materials, but also using air to dry them was developed, in the Building Physics Laboratory at the University of Porto-Faculty of Engineering a new treatment technique called the "Wall Base Ventilation System". The technique consists of circulating dry air around the base of very thick walls built from different materials. Wall base ventilation increases evaporation, which reduces the level of the damp front.

In the following sections two examples of the application of the "Wall Base Ventilation System" as an air drying technology to treat pathologies observed in buildings are presented.

## 1.3 Relative Humidity Control in Crawl Spaces

Mould growth has been found on many different types of materials. Most common ones are organic materials such as wood and materials with organic components such as gypsum boards with paper surfaces. The mould growth on wooden

**Fig. 1.1** The museum house

material has been the subject of experimental research for a long time, with the aim to describe the material response and find the critical conditions for mould growth on surfaces of different materials (Block 1953; Henningsson 1980; Park 1982).

The wood rotting process is divided in two phases (Nofal and Kumaran 2000). In the first phase, called the "initial stage", fungi in the air stick to the wood surface and their hypha grow into the pores in the wood. When the fungi have grown to a certain size, they start the so calling "eat wood". This is the beginning of the second phase, entitled the "growth stage" (Iwamae and Matsumoto 2003; Iwamae and Suzuki 2008).

The most important factors causing mould growth in a crawl space are relative humidity, temperature, nutrition and pH. In practice, relative humidity is the most significant factor causing mould growth in the crawl spaces. The limit value for relative humidity in crawl spaces is usually considered to be from 75 to 80 % (Samuelsson 1994). Relative humidity and temperature are strongly linked together. The thermal mass of the structures and ground soil affects the temperature behaviour of the crawl space.

In this example, a Portuguese museum house, located near Porto, with wood rot problems over the crawl space, is presented. After the museum house rehabilitation in October 2008, part of the building noticed moisture damage on the wood-floor over the crawl spaces (see Fig. 1.1). It is well-known that wood-floors of historic buildings, when they are unventilated correctly, often face problems of biological degradation.

After the rehabilitation in 2008, the ventilation grills were partially closed and the crawl space was poor ventilation, essential to avoid moisture damage. In literature it is possible to find different standard building codes specifies with different ratios of foundation ventilation area and crawl space area. Two years later the rehabilitation the museum house presents several pathologies (see Fig. 1.2a, b) as:

- Wood floor with deformations and without strength;
- Micro-organisms and fungal development in the crawl space with high values of relative humidity.

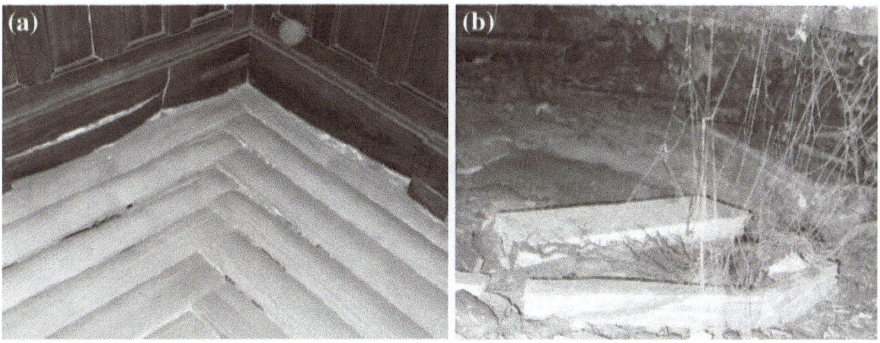


**Fig. 1.2** **a** Wood floor with deformations and without strength, and **b** micro-organisms and fungal development in the crawl space

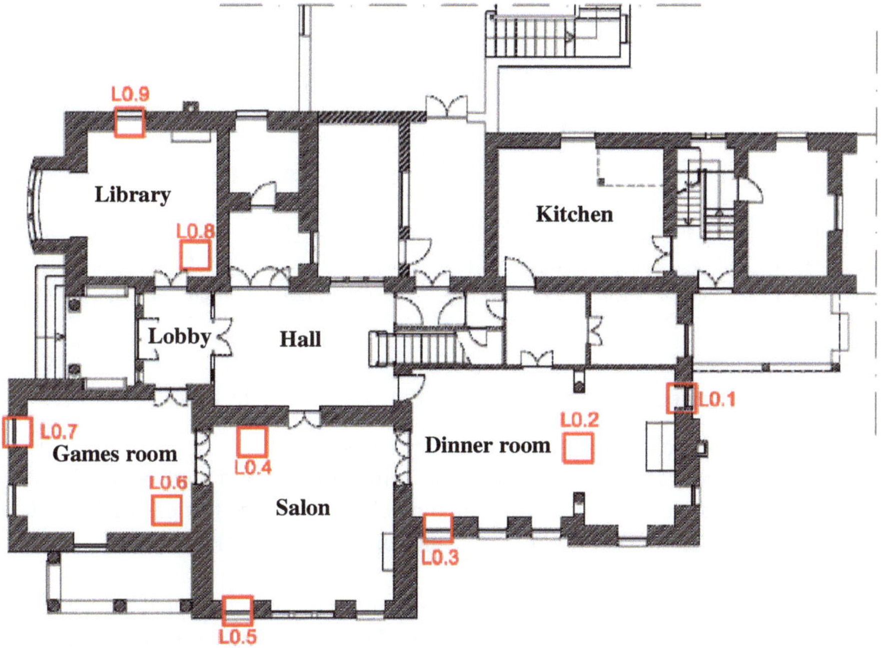


**Fig. 1.3** Data-loggers location on crawl spaces

The pathology observed were deformed floor surface without mechanical resistance, humidity, biological degradation and development of micro-organisms and fungi.

An experimental campaign was carried out with continuous monitorised of the relative humidity and temperature on the crawl spaces below games room, salon, dinner room and library (see Fig. 1.3).

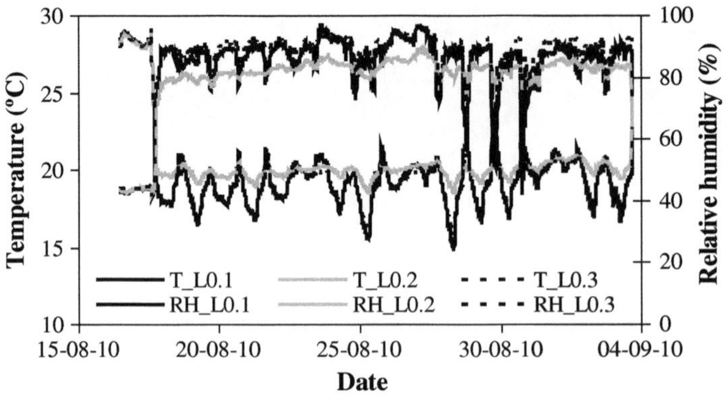


**Fig. 1.4** Example of interior relative humidity and temperature (L0.1, L0.2 and L0.3)

**Table 1.1** Maximum, minimum and average values of temperature and relative humidity measured by the different data-loggers

| | T (°C) | RH (%) | T (°C) | RH (%) | T (°C) | RH (%) | T (°C) | RH (%) | T (°C) | RH (%) |
|---|---|---|---|---|---|---|---|---|---|---|
| | *Outside* | | *Inside* | | *L0.1* | | *L0.2* | | *L0.4* | |
| Mean | 21.8 | 70.7 | 22.5 | 66.9 | 20.5 | 79.6 | 20.5 | 79.6 | 21.4 | 87.7 |
| Min. | 15.3 | 16.7 | 21.4 | 59.0 | 18.5 | 41.2 | 18.5 | 41.2 | 20.3 | 40.9 |
| Max. | 37.6 | 100 | 24.5 | 76.0 | 28.8 | 89.3 | 28.8 | 89.3 | 29.5 | 95.6 |
| | *L0.5* | | *L0.6* | | *L0.7* | | *L0.8* | | *L0.9* | |
| Mean | 20.3 | 82.2 | 22.2 | 79.1 | 22.2 | 79.1 | 21.8 | 86.2 | 22.3 | 76.3 |
| Min. | 14.5 | 41.8 | 16.9 | 41.4 | 16.9 | 41.4 | 20.1 | 41.4 | 16.6 | 41.5 |
| Max. | 28.7 | 95.0 | 28.9 | 93.7 | 28.9 | 93.7 | 29.0 | 93.9 | 29.5 | 92.2 |

The results monotorised show that (see Fig. 1.4 and Table 1.1):

- Inside and outside temperatures were very similar;
- Relative humidity in the crawl space below dinner room was approximately 80 %;
- In the crawl space below the salon the relative humidity measured by data-logger L0.4 was approximately 90 % and outside mean RH was lower than the inside mean RH;
- Relative humidity in the crawl space below games room was approximately 93 %;
- Relative humidity in the crawl space below library was approximately 86 %.

### 1.3.1 Prediction of Mould Growth

In order to predict mould growth, the equation developed by Viitanen (1996), for wood (pine and spruce), was used,

$$t = \exp\left(-0.68\ \ln T - 13.9\ \ln RH + 0.14W - 0.33S + 66.02\right) \tag{1.1}$$

**Table 1.2** Values of time for mould growth, time of wetness and mould growth index obtained in the museum house

| | L0.2 | L0.4 | L0.6 | L0.8 |
|---|---|---|---|---|
| TOW (weeks) | 2.39 | 2.41 | 2.41 | 2.51 |
| $t$ (weeks) | 10.26 | 2.50 | 1.01 | 2.83 |
| $M$ | 0.23 | 0.96 | 2.42 | 0.83 |

where $t$ is the time for mould growth to reach the microscopic stage (weeks), $T$ the temperature (0.1–40 °C), $RH$ the relative humidity (%), $W$ the wood species (pine = 0, spruce = 1) and $S$ is the factor describing the nutrients on the wooden surface (a re-sawed surface after drying = 0, a kiln-dried surface = 1).

The criterion to avoid wood mould growth is that $t$ must be shorter than the time of wetness (TOW), i.e. there is no risk of mould growth if

$$\text{TOW} \ll t \tag{1.2}$$

TOW is defined here as the cumulative time period when $RH$ is higher than 75 %. This means that the counting of TOW starts when $RH$ exceeds 75 %. The periods when $RH < 75$ % are not taken into account. The calculation ends when the duration of the dry period ($RH < 75$ %) is longer than 2 months (this is the time needed to reset the contamination). The temperature and $RH$ that should be used for the calculation of TOW are average values of 2 weeks period. This period is chosen since it is the shortest possible time needed for mould growth to start, and long enough to perform manual calculations of TOW.

In our case study, the temperature and $RH$ used for the calculation of TOW are average values of period in analyse (20 days). In Eq. (1.1), average values of temperature and RH of the cumulative TOW period are used. In this work, we have pine wood ($W = 0$) and the surface has not been sawed after drying ($S = 1$).

For varying temperature and humidity conditions, Viitanen (1996) interprets Eq. (1.1) as a differential equation

$$\frac{dM}{dt} = \frac{1}{\exp\left(-0.68 \ln T - 13.9 \ln RH + 0.14W - 0.33S + 66.02\right)} \tag{1.3}$$

where $M$ is the mould growth index and $t$ is the time calculated in weeks. The mould growth index illustrates the growth rate. The scale for $M$ is: $M \approx 0$ no growth is observed, $M \approx 1$ some growth is detected only with microscope, $M \approx 3$ some growth is detected visually and $M \approx 6$ very heavy and dense mould growths covers nearly 100 % of the surface.

The results presented in Table 1.2 show a high prediction of mould growth on games room (L0.6), TOW $> t$ and $M \approx 3$, and significant possibility of mould growth on salon (L0.4) and library (L0.8), TOW $< t$ and $M \approx 1$. The only room analysed with lower prediction of mould growth, for the period studied, is the dinner room. These results were confirmed by in situ observations.

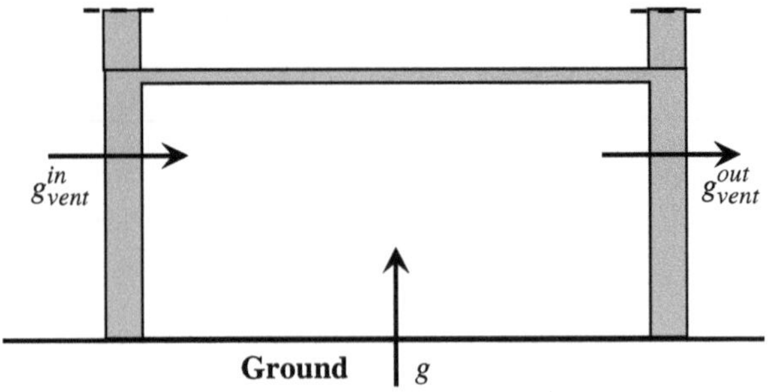


**Fig. 1.5** Sketch of modelled moisture ($g$) flow in crawl space

## *1.3.2 Moisture Balance in Crawl Spaces*

In the crawl space, it is usually considered that a relative humidity over 75–80 % (Samuelsson 1994), during a period of several weeks or months, can cause mould growth. Temperature is high enough for mould growth because it is usually over +5 °C in crawl spaces. Relative humidity in the crawl space is the result of ground moisture evaporation, air change rate and thermal behaviour that are all strongly linked.

As sketched in Fig. 1.5, in a general state the humidity balance for the air in the crawl space is given by (Elmroth 1975; Kurnitski 2001):

$$V\frac{\partial C_{air}}{\partial t} = g_{vent}^{in} + g - g_{vent}^{out} \tag{1.4}$$

where $V$ is the volume of crawl space (m$^3$), $g$ is the moisture flow (kg/s) and $C_{air}$ is the humidity by volume of crawl space air (kg/m$^3$).

If the ground moisture evaporation and moisture flows carried by air change are the only moisture flows in the crawl space, if moisture storage and flows in constructions are not considered, and rain and surface water are not present in the crawl space as should be provided by correct design. Therefore, the moisture balance can be written in steady state

$$g_{vent}^{out} = g_{vent}^{in} + g \tag{1.5}$$

or

$$C_{air}^{out} q_m = C_{air} q_m + g \tag{1.6}$$

where $C_{air}^{out}$ is the outdoor air humidity by volume (kg/m$^3$), $q_m$ is the air change in the crawl space (m$^3$/s) and $C_{air}$ is the humidity by volume in extract air (considering that with complete mixing it is the same everywhere in the crawl space air), (kg/m$^3$). Ground moisture evaporation based on mass transfer coefficient and potential is

$$g = \beta\left(p_{ground} - p_{air}\right)\frac{M_w}{R \cdot T} A_{evap} \tag{1.7}$$

where $\beta$ is the mass transfer coefficient (m/s), $p_{ground}$ is the vapour pressure on the ground surface (Pa) and $p_{air}$ in the crawl space air (Pa), $M_w$ is the molecular weight of water (0.018 kg/mol), $R$ is the universal gas constant (8.31 J/mol K), $T$ the absolute temperature (K) and $A_{evap}$ is the evaporation surface area (m$^2$). If the ground soil is relatively dry, Eq. (1.7) overestimates the evaporation rate and alternatively Fick's law can be used for moisture transfer inside the ground soil.

To determine the mass transfer coefficient, $\beta$, the convective heat transfer coefficient could be used

$$\beta = \frac{\alpha}{\rho \cdot c_p}\frac{\rho}{\rho_{BM}}Le^{1-n} \approx \frac{\alpha}{\rho \cdot c_p} \approx \frac{2.2\left(T_{ground} - T_{air}\right)^{1/3}}{\rho \cdot c_p} \tag{1.8}$$

where $\alpha$ is the convective heat transfer coefficient (W/m$^2$ K), $\rho$ is the density of air (kg/m$^3$), $c_p$ is the specific heat capacity of air (J/kg K), $\rho/\rho_{BM}$ is the logarithmic density term, $Le$ is the Lewis number, $T_{ground}$ is the temperature on the ground surface and $T_{air}$ is the temperature on crawl space air. The last term of Eq. (1.8) is the most common expression for mass transfer coefficients and gives a sufficiently accurate result in the crawl space with average air change until 1.5 ach, assumption of laminar boundary layer. For high values of average air change the following equation is more appropriate

$$\beta \approx \frac{2.2\left(T_{ground} - T_{air}\right)^{1/3} + 4v}{\rho \cdot c_p} \tag{1.9}$$

where $v$ is the air velocity (m/s) stated as a linear function of air change: $4v = 5.4q$, where $q$ is the airflow (m$^3$/s). Finally, if the ground soil is relatively dry, Eq. (1.8) overestimates the evaporation rate and alternatively the Fick's law should be used for moisture transfer inside the ground surface.

In our case study, for the time period in analyse (1 month) and using Eq. (1.6) the average value of moisture evaporation was approximately 0.21 g/m$^2$h (natural ventilation of 0.1 ach), corresponding to a mass transfer coefficient, $\beta$, of approximately $6.6 \times 10^{-5}$ m/s.

The development of Eq. (1.4) results in

$$V\frac{\partial C_{air}}{\partial t} = achV\left(C_{air}^{out} - C_{air}\right) + \beta\frac{V}{L}\left(C_{ground} - C_{air}\right) \tag{1.10}$$

where $L$ is the height of the crawl space. The integration of Eq. (1.10) for the initial and end conditions $t = 0$ to $t = t$ and $C = C_0$ to $C = C$, gives

$$C_{air} = \frac{achC_{air}^{out} + \frac{\beta}{L}C_{ground}}{ach + \frac{\beta}{L}} - \frac{e^{-t\left(ach+\frac{\beta}{L}\right)+\ln\left[ach(C_{air}^{out}-C_0)+\frac{\beta}{L}(C_{ground}-C_0)\right]}}{ach + \frac{\beta}{L}} \tag{1.11}$$

For $t \to \infty$, steady-state, Eq. (1.11) results in

$$C_{air} = \frac{achC_{air}^{out} + \frac{\beta}{L}C_{ground}}{ach + \frac{\beta}{L}} \tag{1.12}$$

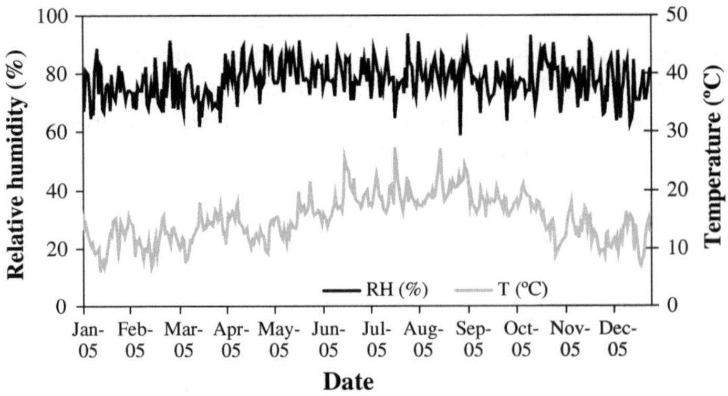


**Fig. 1.6** Daily average values of exterior relative humidity and temperature

So, for the games room crawl space ($A_{evap} = 40\,\text{m}^2$ and $L = 0.5$ m) with an $ach = 0.1\,\text{h}^{-1}$ and considering an average value of temperature and relative humidity of 15 °C and 77 % (see Fig. 1.6), for Porto, and $C_{ground} = 0.013\,\text{kg/m}^3$ (corresponding a value of relative humidity of 97 %, practically fully saturated), results by Eq. (1.12) a $C_{air}$ of 0.012 kg/m$^3$, corresponding a value of relative humidity of 94 %. Increasing the $ach$ value to 0.5 h$^{-1}$ (10 m$^3$/h) and 2.5 h$^{-1}$ (50 m$^3$/h), results by Eq. (1.12) a relative humidity value on the crawl space of 87 and 80 %, respectively.

It is possible to observe that the value obtained with Eq. (1.12), $C_{air} = 0.012\,\text{kg/m}^3$ and $RH = 94\,\%$, for the case of crawl space with natural ventilation, is practically the same obtained in in situ experiments, $RH = 93\,\%$ (L0.6). The results, also, show that the use of natural ventilation on this crawl space is insufficiently to avoid mould growth ($RH < 80\,\%$).

### *1.3.3 Numerical Simulation*

There are an increasing number of computer models that will model heat and mass transfer in building materials (Hens 1996). The calculation program used in the numerical simulations for the experimental and analytical validations was WUFI-2D, developed by the Fraunhofer Institute for Building Physics. The governing equation for moisture transport is (Holm and Kunzel 2003)

$$\frac{dw}{d\phi}\frac{\partial\phi}{\partial t} = \nabla\left(D_{\varphi}\nabla\phi + \delta\nabla\phi p_a\right) \tag{1.13}$$

and the governing equation for heat transport is

$$\frac{dH}{dT}\frac{\partial T}{\partial t} = \nabla(\lambda_{\text{T}}\nabla T) + h_{\text{v}}\nabla(\delta_{\text{p}}\nabla\phi p_{\text{a}}) \tag{1.14}$$

where $dH/dT$ is the heat storage capacity of the moist building material, $dw/d\phi$ is the moisture storage capacity, $\lambda_T$ is the thermal conductivity, $D_\phi$ is the liquid conduction coefficient, $\delta_p$ is the water vapour permeability, $h_v$ is the evaporation enthalpy of the water, $p_a$ is the water vapour saturation pressure, $T$ is the temperature and $\phi$ is the relative humidity (Evrard and De Herde 2010).

In the numerical analysis used to validate the analytical study only the change in relative humidity was considered important, since the experiments took place under isothermal conditions.

The time of calculation lasted from January, 1st to December, 31st of the year 2005. The simulation was carried out in time steps of 5 min to reproduce a realistic control. Given that the location of the museum house is in Estarreja, which is located near Porto, the meteorological data of Porto for the year 2005 was used for the simulation, as exterior climate. A weather file was created which consisted of 5 min values of exterior temperature, relative humidity, global and diffuse radiation, rain, wind and barometric pressure. The interior climate considered was a sinusoidal function with an average value of temperature and relative humidity of 22 ± 3 °C and 65 ± 20 %, respectively. The short wave radiation absorptivity and long-wave radiation emissivity considered were 0.4 (stucco-normal bright) and 0.9, respectively, and the initial conditions within the element were $\phi = 70$ % and $T = 20$ °C.

The vapour diffusion thickness value used was zero (no coating) and the interior heat transfer coefficient was constant and equal to 8 $W/m^2K$. The exterior heat transfer coefficient only contained the convective element and was considered independent of wind (at a constant value of 17 $W/m^2K$).

The simulations used the real climatic variables of the city of Porto and actual material properties of the museum house (see Table 1.3). Detailed model geometry of the games room was created in WUFI-2D (see Fig. 1.7) to be used as an example of numerical study.

In this case study simulations were done, for 1 year, with exterior climate of Porto (see Fig. 1.6) to analyse the influence of ventilation in a crawl space with wet ground, practically fully saturated (≈100 % RH).

Figures 1.8, 1.9 and 1.10 show the importance of good ventilation in crawl spaces. In Fig. 1.8 it is possible to observe the relative humidity on crawl space near the wood-floor and in the middle of the wood-floor. This figure shows an extremely high tendency, of a crawl space in a moderate climate (Porto) without ventilation, to present mould growth (during February to August).

In Figs. 1.9 and 1.10 the numerical simulations results for the same crawl space but with ventilation, $ach = 0.1\ h^{-1}$ (Fig. 1.9) and $ach = 2.5\ h^{-1}$ (Fig. 1.10) are presented. Figure 1.9 shows that reduced ventilation of the crawl space implies a high tendency to present mould growth. The use of mechanical ventilation with an *ach* of 2.5 $h^{-1}$ (Fig. 1.10) decreases the relative humidity but do not solve the problem of mould growth.

In conclusion, this case study show that the ventilation is needed to avoid mould growth in crawl spaces. However, it is preferable to use controlled mechanical ventilation, a hygro-regulable mechanical ventilation device, that

**Table 1.3** Material properties of Limestone

| | | | |
|---|---|---|---|
| Bulk density, $\rho$ (kg/m$^3$) | | | 2,155 |
| Heat capacity, $c_p$ (J/kgK) | | | 1,000 |
| Porosity, $\varepsilon$ (%) | | | 19.7 |
| Thermal conductivity, $\lambda_T$ (W/mK) | | | 1.33 |
| Vapour diffusion resistance factor, $\mu$ (–) | Dry cup | | 41 |
| | Wet cup | | 29 |
| Moisture storage function, $w$ (kg/m$^3$) | $\phi$ (%) | *Adsor* | *Desadsor* |
| | 4.0 | 0.521 | 0.951 |
| | 11.2 | 0.593 | 1.150 |
| | 34.8 | 0.872 | 1.239 |
| | 58.6 | 1.043 | 1.628 |
| | 76.3 | 1.237 | 2.360 |
| | 80.0 | 1.334 | – |
| | 84.2 | 1.584 | – |
| | 92.1 | 2.831 | 4.144 |
| Capillary transport coefficient, $D_w$ (m$^2$/s) | $w = 1.7$ kg/m$^3$ | | $6.6 \times 10^{-11}$ |
| | $w = 188$ kg/m$^3$ | | $6.2 \times 10^{-8}$ |
| Water absorption coefficient, $A$ (kg/m$^2$ s$^{1/2}$) | | | 0.024 |
| Free-water saturation, $w_f$ (kg/m$^3$) | | | 177 |

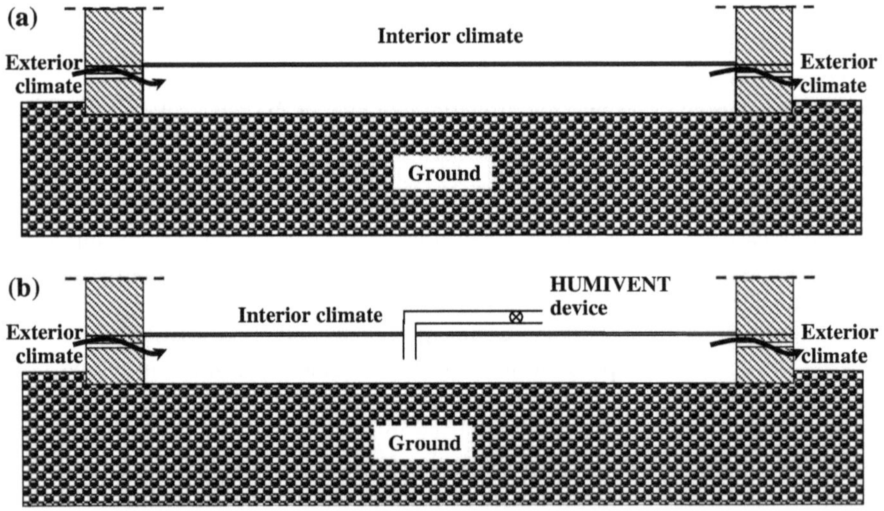


**Fig. 1.7** Detailed geometry, used with WUFI-2D, of the analysed building area (games room): **a** without the hygro-regulated system, and **b** with the hygro-regulated system

coming into operation only when the outside water vapour pressure is lower than the partial pressure inside the crawl space. We believe that a hygro-regulable system it is a good solution for this kind of mould problems in crawl spaces (Hagentoft et al. 2008).

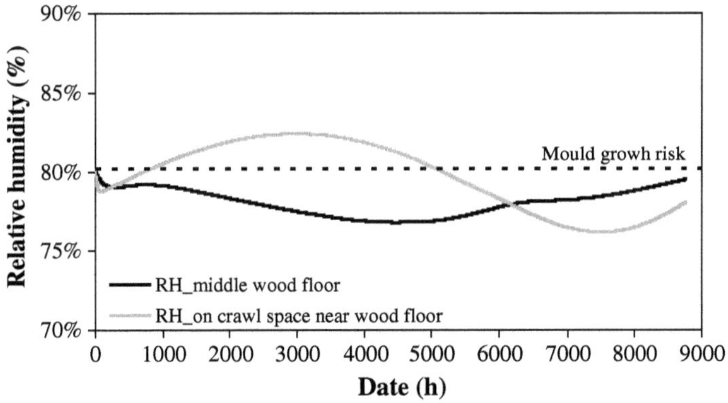


**Fig. 1.8** Values of relative humidity in a crawl space without ventilation grill

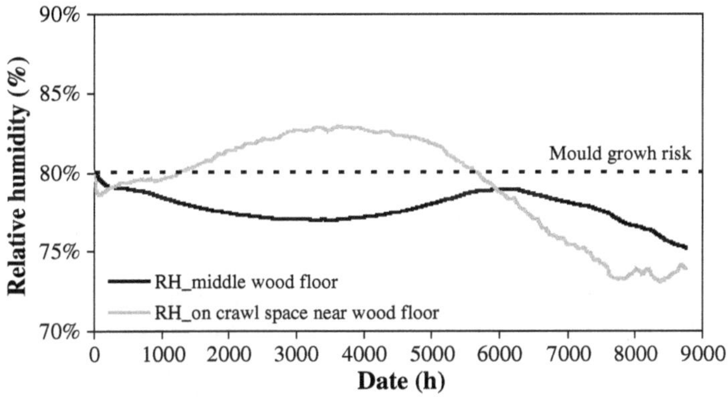


**Fig. 1.9** Values of relative humidity in a crawl space with small ventilation ($ach = 0.1\ \text{h}^{-1}$)

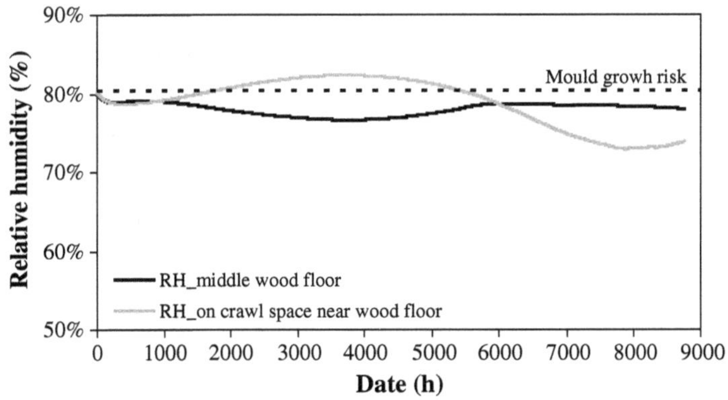


**Fig. 1.10** Values of relative humidity in a crawl space with ventilation ($ach = 2.5\ \text{h}^{-1}$)

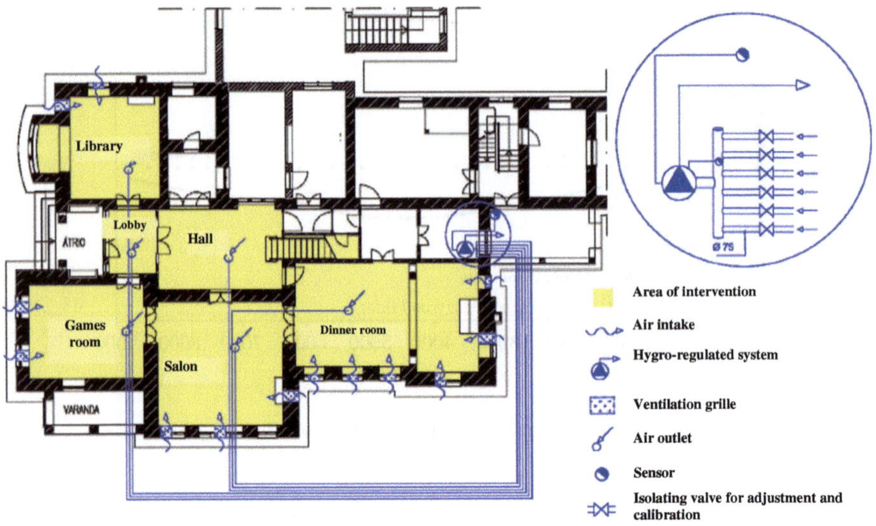


**Fig. 1.11** Scheme of the intervention methodology proposed to the crawl spaces

## *1.3.4 Proposed Intervention Methodology: The Hygro-regulated System*

Reducing the moisture level in the air (relative humidity) and maintaining the inside relative humidity below 75 % it is the key to control mould growth. So, the intervention methodology proposed could be divided in two steps: control the groundwater level and ventilating the crawl spaces with a hygro-regulated system (Freitas et al. 2011; Guimarães et al. 2010).

The procedure adopted for the crawl spaces was the implementation of a hygro-regulated ventilation system with the following characteristics:

- Creating a ventilation network, by crawl space module, consisting of a pipe with 0.075 m of diameter associated with a hygro-regulated system, developed by FEUP, University of Porto (see Fig. 1.11);
- The ventilation pipes should be individual, with different flow regulations, and all the pipes converge in a single pipe, where the temperature and relative humidity of the crawl space air is measured;
- The hygro-regulated device receive records related to the inside and outside hygrothermal conditions, and coming into operation only when the outside water vapour pressure is lower than the partial pressure inside of the crawl space;
- The ventilator should have variable speed with flows between 10 and 50 $m^3$/h.

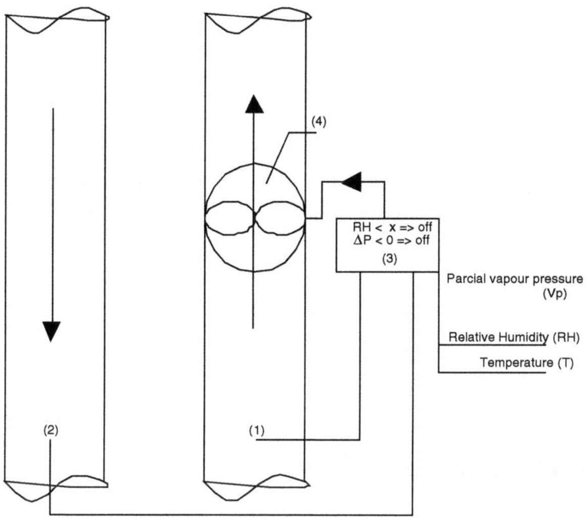


**Legend:**
Probe (1) – Entrance (temperature and relative humidity)
Probe (2) – Exit (temperature and relative humidity)
Fan with variable speed (3)
Control module (4):
Probe 1 – $T_1$ e $RH_1$ => Vapour pressure 1 (Pa)
Probe 2 – $T_2$ e $RH_2$ => Vapour pressure 2 (Pa)
Calculate $\Delta P = P_2 - P_1$
Instructions of ventilator operation
ON if $\Delta P \geq 0$ and $RH_1$ > Predetermined value
OFF if $\Delta P \leq 0$ or $RH_1$ < Predetermined value

**Fig. 1.12** The principle of functioning of a hygro-regulated wall base ventilation system (Freitas et al. 2011)

In this novel device, the extraction is controlled by a variable ventilator, hygro-regulated, which comes into operation when the inside water vapour pressure is higher than the outside water vapour pressure, and when the inside relative humidity is higher than a certain predefined value to guarantee no salt crystallization / dissolution problems inside the system (see Fig. 1.12).

The control module receives information from two probes (inside and outside temperature and relative humidity), calculates the inside and outside water vapour pressure and evaluates the positive or negative sign of the pressure differential to shutting on or down the ventilator.

The prototype produced and tested is composed by a programmer data-logger, linked to two relative humidity (*RH*) and temperature (*T*) probes, link up to one to four ventilators at the same time, a battery, a connection to a computer to make the operation device programming and giving instructions to the data-logger to get temporal *RH* and *T* information, reading of records and data processing (see Fig. 1.13).

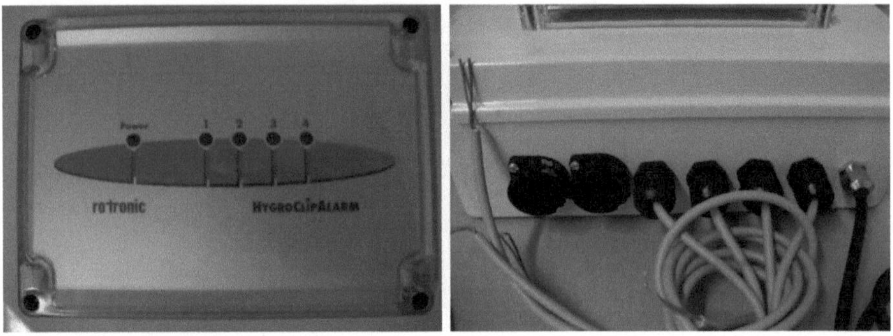


**Fig. 1.13** The prototype

## 1.4 Rising Damp in Walls

Rising damp, which originates in the ground and permeates porous materials by capillarity, is one of the main causes of decay in old buildings, particularly those containing thick walls built from different materials. Treating rising damp in the walls of historical buildings is a very complex procedure. Moisture transfer in the walls of historical buildings in direct contact with the ground leads to the migration of soluble salts, which are responsible for many building pathologies (Anderberg and Wadsö 2007; Karoglou et al. 2005).

The mechanisms describing the moisture transport in building walls are complex. During the vapour phase, diffusion and convection play the main part, while in the liquid phase the transfer is controlled by the capillary action, gravity and the pressure gradient effect (Torres and Freitas 2007). In practice, the transport occurs in the liquid and vapour phases simultaneously, and is dependent upon environmental conditions such as temperature, relative humidity, precipitation, solar radiation and atmospheric wind pressure (which define the boundary conditions) and the characteristics of the building materials used. From the physical point of view, there are three main mechanisms involved in moisture fixation: hygroscopicity, condensation and capillarity. In most cases, these three mechanisms are responsible for moisture content variation of porous building materials with a porous structure.

Having identified the particular characteristics of those buildings, it is important to recognise the limitations of traditional technologies and to investigate new solutions for the phenomenon. Rising damp may usually be controlled by adopting one or more of the following treatment techniques as: physical or chemical barrier, by reducing the absorbent section (Massari and Massari 1985), by watertight barriers (Massari 1971) or by applying products (BS 6576 2005; L'Anson and Hoff 1988); another traditional techniques are using a potential against the capillary potential (Bertollini et al. 2009), an atmospheric drainage (Henriques 1993), applying a coating with controlled porosity (Colombert 1975), concealing anomalies (CSTB 1985) and ventilating the wall (Heiman 1982). The traditional

**Fig. 1.14** Building pathologies observed in the Portuguese historical church

techniques used to deal with this kind of problem sometimes prove ineffective or are very expensive, in particular when dealing with walls of considerable thickness and heterogeneous materials (Torres and Freitas 2007; Guimarães et al. 2010; Torres and Freitas 2010).

This example presents the application of the wall base ventilation system to treat rising damp phenomena in a historical church located in Northern Portugal. The church presented several pathologies (see Fig. 1.14) as: green stains, fungal developments, surface condensations and damp patches.

The hygro-regulable system was installed in a historical church in Northern Portugal, with thick walls and heterogeneous composition (see Fig. 1.15a). Inside the building, two hygro-regulable mechanical ventilation subsystems were installed (the first in a north facing wall with a length of 122 m and the second in a south facing wall, with a length of 45 m). In the Southside subsystem, the air was admitted into the system through grids located inside the building (indoor air) and was extracted into the cloister (see Fig. 1.15b). In the Northern subsystem, the air was admitted from the outside (outdoor air).

Extraction was controlled by a hygro-regulable engine of variable speed (Freitas and Guimarães 2008). The probes were installed, one at the inlet and the other outlet the system. Each probe has a transmitter that sends the results (relative humidity and temperature) to a data logger. The control module gives orders to the ventilator device to turn on or to turn off the system according to the criterion of functioning. This device transforms the system into a hygro-regulable mechanical ventilation.

The system installed in the church initially began operating whenever the relative humidity at the outlet was 5 % higher than the relative humidity at the inlet

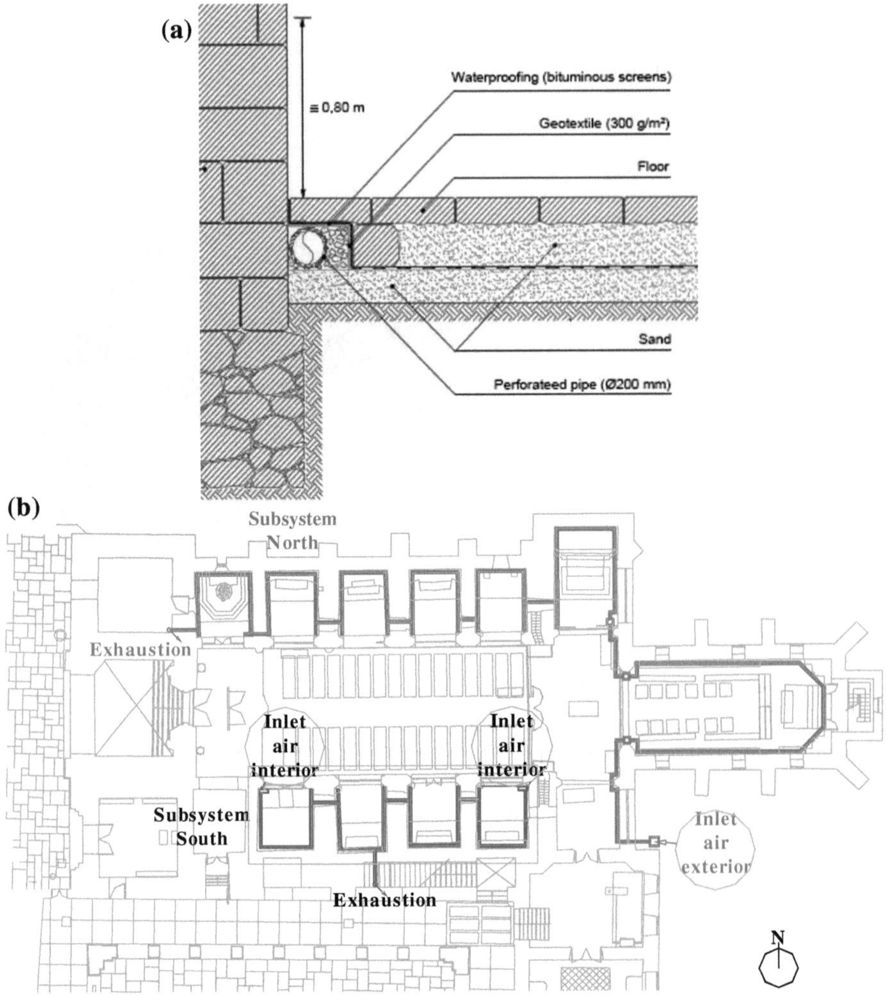


**Fig. 1.15** **a** The wall base ventilation system used in the historical church, and **b** northern and a southern subsystems

(Torres 2004). The idea was to admit dry air as compared to the air inside the system. This criterion was found to be inadequate, as it meant that the system was operating at periods when condensation occurred inside it. The inlet vapour pressure was sometimes higher than the outlet vapour pressure flux. Consequently, a new criterion was proposed with a view to optimizing the system, based upon the difference in vapour pressure ($\Delta P$) at the outlet and inlet. The system now began functioning whenever the $\Delta P$ was positive. In order to avoid the occurrence of salts crystallizations, and/or dissolutions, another criterion was added, namely, to limit the inlet value of relative humidity (see Fig. 1.12).

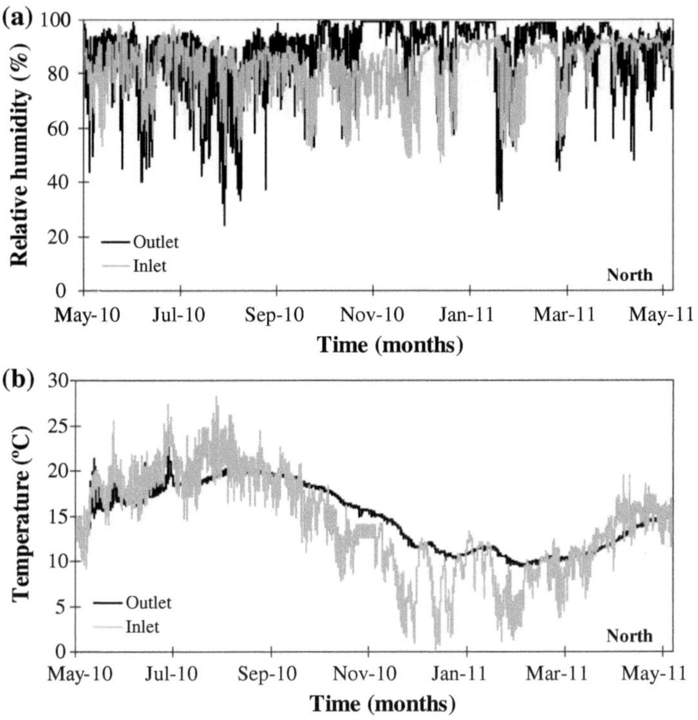


**Fig. 1.16** Relative humidity (**a**), and temperature (**b**) variation of the northern subsystem

## *1.4.1 Experimental Results*

Figures 1.16 and 1.17 show, respectively, the relative humidity and temperature differences at the inlet and at the outlet of the subsystems. These values were directly read from the data-logger installed.

Figures 1.16a and 1.17a show that the inlet relative humidity of these 2 subsystems presents high values throughout the year, with an average value close to 80 % in the South and close to 85 % (sometimes 100 %) in the North. These values in conjunction with the high temperatures of the air admitted in the summer contributed to the occurrence of condensations.

During summer months the average inlet temperature was higher than the outlet temperature at the exit. On the other hand, during the winter months the average inlet temperature was lower than the temperature at the exit (see Figs. 1.16b and 1.17b). The values of the temperature inside the system, next to the outlet, followed up on the values of the inlet air temperature, although with a clear gap, especially in the southern subsystem. The thermal inertia is an important factor that justifies this behaviour.

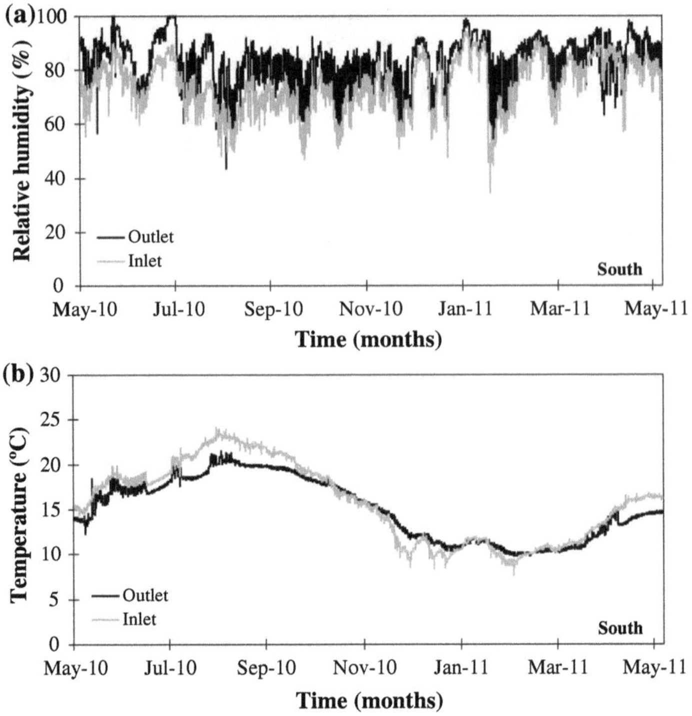


**Fig. 1.17** Relative humidity (**a**), and temperature (**b**) variation of the southern subsystem

Figure 1.18a, b shows the vapour pressure differences between the inlet and outlet subsystems, and the daily average values to better assess their behaviour, in particular, the possibility of condensations inside the subsystems.

### *1.4.2 Seasonal Operation System*

Figure 1.19a, b shows the functioning periods of the ventilator, in a northern and a southern wall, and the importance of the wall base ventilation system in the treatment of rising damp, by the quantity of water vapour extracted during the 5 months analysed (386.9 kg in the north wall and 454.6 kg in the south wall). These figures show that the operation periods of the southern subsystem occurred mainly during the winter months with longer periods of inactivity during the summer months. On the other hand, the northern subsystem presents a high number of interruptions distributed throughout the year. In this subsystem the air was admitted from the outside of the Church and, as a result the vapour pressure values were less stable (sometimes with very abrupt variations), which also could justify the high number of fan interruptions.

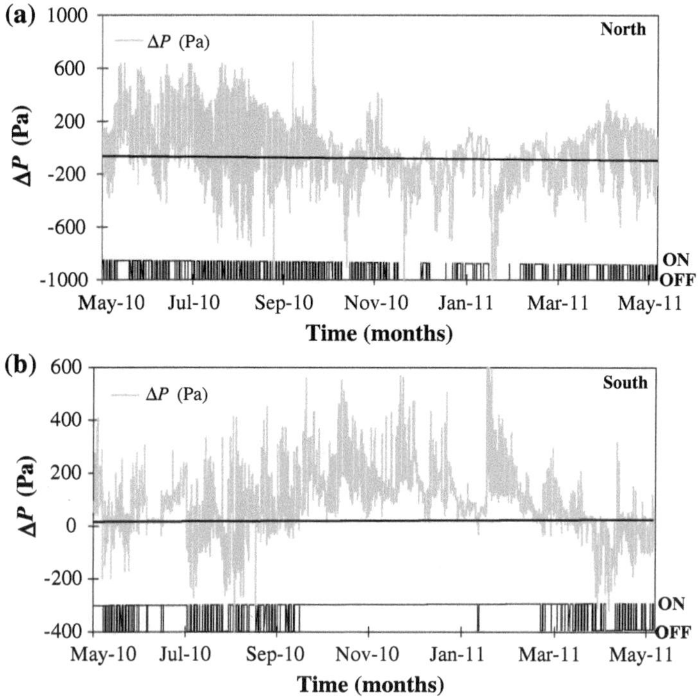


**Fig. 1.18** Vapour pressure differential and functioning of system: **a** north wall, and **b** south wall

### *1.4.3 Systems Optimization*

The southern subsystem presented a more effective behaviour during the winter months, where when it removed more water than the northern subsystem. In contrast, during the summer months, the northern subsystem was more effective than the southern subsystem. However, it is important to have in mind that the properties of the air inlet were very different in the two subsystems. The southern subsystem admitted air coming from inside the church and the northern subsystem admitted air from the outside. So, for this case study, a church located in the north of Portugal, the wall base ventilation system should work with a seasonal optimised functioning, i.e., during the summer months (between April and September) the system should only use outside air and during the winter months (between October and March) it should use air coming from inside the church.

For a better performance or efficiency of the wall base ventilation system installed it is important to reduce the interruption periods when potential condensations occur inside the system. A continuous functioning of the wall base ventilation system with a constant evaporation results in a greater amount of water withdrawn. In accordance with the optimised scenario, the air is admitted from the outside, between April and September, and from the inside between

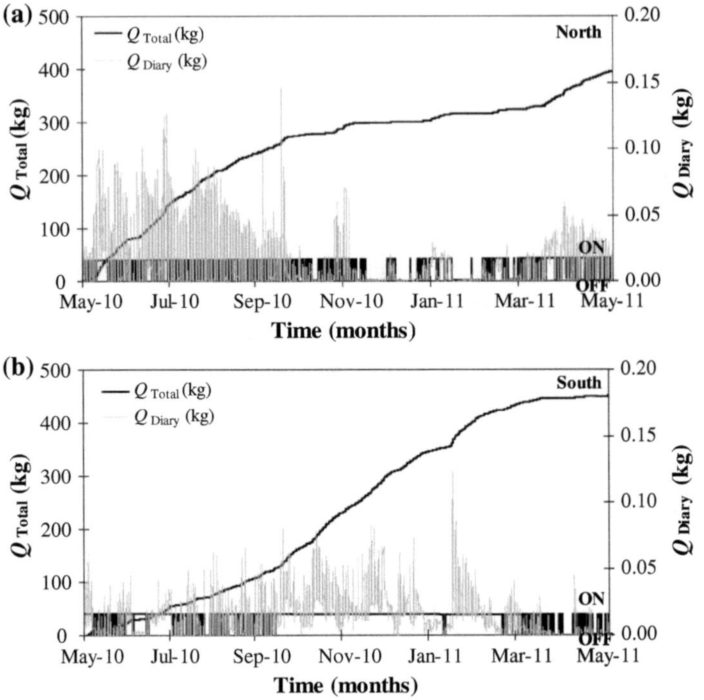


**Fig. 1.19** Quantity of water vapour extracted from system: **a** north wall, and **b** south wall

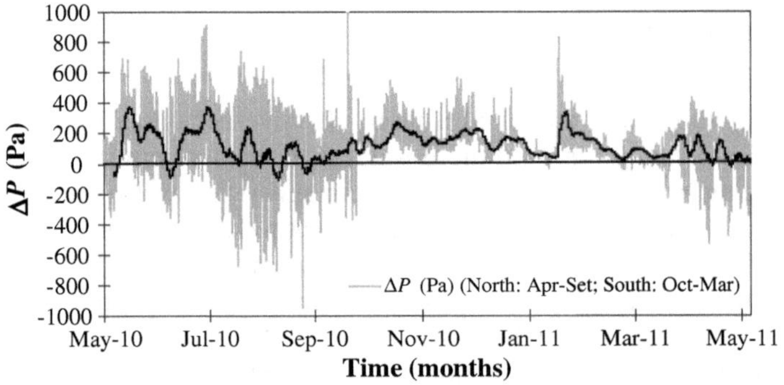


**Fig. 1.20** Water vapour pressure variation and subsystems operation periods

October and March. Figure 1.20 shows the values of the water vapour pressure difference between the outside and the inside. In this figure it is possible to observe the occurrence of some condensations, during the summer months, in periods when the system is not working. However, since it is an open system, with natural ventilation during the periods of inactivity, this type of condensation may, occasionally, occur when the air comes from the outside with abrupt

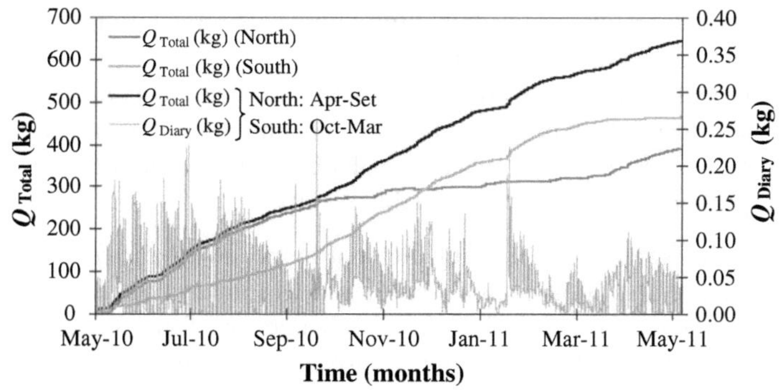


**Fig. 1.21** Amount of water and accumulated amount of water withdrawn by the subsystems with a seasonal optimised systems functioning

variations of relative humidity and temperature. This can be controlled by shutting down the system each time it stops, in order to bring the ventilation flow to null and thus ensure an effective interruption of the air circulation.

Figure 1.21 shows the prediction of the amount of water and accumulated amount of water withdrawn during 1 year, with the systems functioning in seasonal optimised operation. In this figure it is possible to observe that the amount of water withdrawn in this optimal situation would be much higher than the one withdrawn in either of the two subsystems installed.

Our research group is currently studying whether these values may have adverse consequences in terms of the long-term deterioration of the wall materials (Watt and Colston 2000). The entry of air with very low relative humidity could generate the crystallization of salts that can be found in the building materials, threatening their durability. For this reason, the relative humidity value at the entrance had to be limited. The relative humidity scores recorded, which range from 60 to 95 %, are not considered to present a risk of salt crystallization/dissolution inside the system, taking into account the salts detected in those area. However, the problem might arise in another type of external climate or with other types of salts. In that case the wall base ventilation system must also limit the inlet relative humidity value.

## 1.5 Conclusions

The analysis of the case study related to relative humidity control in crawl spaces resulted in the following conclusions:

- The in situ experimental results showed that high values of relative humidity implies biological degradation of the wood floor;
- The numerical and analytical models used showed the same tendency of the experimental results;

- The numerical results showed the importance of crawl spaces with a good ventilation to avoid mould growth;
- The numerical results, also, suggested that controlled mechanical ventilation is preferable to strongly continuous mechanical ventilation in this type of spaces;
- In numerical analysis of systems that include ground, 1 year of numerical simulation is probably short to obtain reliable results.

Ground moisture evaporation is a key element in moisture balance in crawl spaces. If any moisture evaporation occurs, ventilation will always be required to remove this moisture. However, increasing the air change rate from a certain level only reduce the humidity if the ventilation removes more moisture than is added by increased ground moisture evaporation.

The hygro-regulated device is presented as a good solution to avoid high levels of moisture in crawl spaces. This technique was successfully tested in the treatment of rising damp phenomenon in historical buildings and future experimental results will be used to validate this device in crawl spaces.

The second case study showed the complexity of the mechanisms of moisture transfer, particularly in terms of rising damp in historical buildings. As rising damp is one of the main causes of decay in these buildings, it is important to study the factors related to this phenomenon.

In-field experiments, as in a historical church, show that an efficient way of treating rising damp in historical buildings is to ventilate the wall base. The monitoring results of 1 year showed the values obtained for the relative humidity and temperature at the entry and exit of the northern and southern subsystems, as well as the periods of operation, the vapour pressure differences between the output and the input and the accumulated amount of water vapour carried and withdrawn.

These results suggested a change in the functioning of the two subsystems used: between April and September, the subsystems must admit air from the outside and between October and March they should admit air from the inside. The use of this optimized situation entails a reduction of the interruption periods of the system (with less interior condensations), whose resulting operation will allow for a more permanent and continuous evaporation and a higher amount of water withdrawn.

These two examples of the application of the "Wall Base Ventilation System" as an air drying technology to treat pathologies observed in buildings are presented showed the interest and the viability of the technique.

**Acknowledgments** J. M. P. Q. Delgado would like to thank Fundação para a Ciência e a Tecnologia (FCT) for financial support through the grant SFRH/BPD/84377/2012.

## References

Anderberg A, Wadsö L (2007) Drying and hydration of cement based self-leveling flooring compounds. Drying Technol 25(12):1995–2003

Bertollini L, Coppola L, Gastaldi M, Redaelli E (2009) Electroosmotic transport in porous construction materials and dehumidification of masonry. Constr Build Mater 23(1):254–263

Block SS (1953) Humidity requirements for mould growth. Appl Microbiol 1(6):287–293
BS 6576, British Standards Institution (2005) Code of practice for diagnosis of rising damp in walls of buildings and installation of chemical damp-proof courses, London 2005
Colombert R (1975) The humidity of old buildings; Causes and effects; Diagnosis and solutions. Editions du Moniteur, Paris (in French)
CSTB, Centre Scientifique et Tecnhique du Batiment (1985) The methods of treatment against rising damp masonry. Technical information note 162, Bruxelles, Belgium 1985 (in French)
Delgado JMPQ, Guimarães AS, Freitas VP (2012) The wall base ventilation system applied at different wall geometries—numerical simulation of the evaporative process. Drying Technol 30(1):1–12
Elmroth A (1975) Crawl space foundation. Construction Research, Report 12, Stockholm
Evrard A, De Herde A (2010) Hygrothermal performance of Lime-Hemp wall assemblies. J Building Phys 34(1):5–25
Freitas VP, Guimarães AS, Delgado JMPQ (2011) The HUMIVENT device for rising damp treatment. Recent Patents Eng 5(3):233–240
Freitas VP, Guimarães AS (2008) Characterization of a hygro-regulated wall base ventilation system for treatment of rising damp in historical buildings. In: Proceedings of the 2nd Nordic symposium on building physics, Copenhagen
Guimarães AS, Delgado JMPQ, Freitas VP (2010) Mathematical analysis of the evaporative process of a new technological treatment of rising damp in historic buildings. Build Environ 45(11):2414–2420
Hagentoft CE, Kalagasidis AS, Nilsson SF, Thorin M (2008) Mould growth control in cold attics through adaptive ventilation. In: Proceedings of the 8th Nordic symposium on building physics, Copenhagen, pp 1–8
Heiman JL (1982) An evolution of methods of treating rising damp. In: Lieff M, Trechsel HR (eds) Moisture migration in buildings. ASTM special technical publication 779. American Society for Testing and Materials, pp 121–37
Henningsson B (1980) Thermotolerant moulds on timber during kiln drying. Uppsala, Swedish University of Agricultural Sciences, Department of Forest Products, Note no. 96
Henriques FMA (1993) Humidity in walls—manifestation forms, criteria for quantification and repair solutions analysis. Ph.D. thesis, Universidade Técnica de Lisboa (UTL), Lisbon (in Portuguese)
Hens H (1996) Final report task 1: modelling common exercises, summary reports. Annex 24
Holm AH, Kunzel HM (2003) Two-dimensional transient heat and moisture simulations of rising damp with WUFI-2D. In: Proceedings of the 2nd international conference on building physics, Leuven, 14–18 Sept 2003. Swets & Zeitlinger B.V., Lisse, pp 363–367
Iwamae A, Matsumoto M (2003) The humidity variation in crawl spaces of Japanese houses. J Therm Envelope Build Sci 27(2):123–133
Iwamae A, Suzuki H (2008) Durability of crawl space based on damage due to wood rot. In: Proceedings of the 8th Nordic symposium on building physics, Copenhagen, pp 747–754
Karoglou M, Moropoulou A, Maroulis ZB, Krokida MK (2005) Drying kinetics of some building materials. Drying Technol 23(1–2):305–315
Kurnitski J (2001) Ground moisture evaporation in crawl spaces. Build Environ 36(3):359–373
L'Anson SJ, Hoff WD (1988) Chemical injection remedial treatments for rising damp—I. The interaction of damp-proofing fluids with porous building materials. Build Environ 23(3):171–178
Massari I, Massari G (1985) Damp buildings, old and new. Bull Assoc Preserv Technol 17(1):2–30
Massari G (1971) Buildings in damp and unhealthy conditions. Methods for their sanitation. Eyrolles Edition, Paris
Nofal M, Kumaran K (2000) On implementing experimental biological damage-functions models in durability assessment system. In: Proceedings of Japan-Canada housing R&D experts working group meeting building envelope, pp 111–124

Park D (1982) Phylloplane fungi: tolerance of hyphal tips to drying. Trans British Mycol Soc 79(1):174–179

Samuelsson I (1994) Moisture control in crawl space. ASHRAE Technical Data Bull 10(3):58–64

Torres MIM, Freitas VP (2010) The influence of the thickness of the walls and their properties on the treatment of rising damp in historic buildings. Constr Build Mater 24(8):1331–1339

Torres MIM, Freitas VP (2007) Treatment of rising damp in historical buildings: wall base ventilation. Build Environ 42(1):424–435

Torres MIM (2004) Rising damp in walls of historical buildings. Ph.D. thesis, FCTUC, Coimbra (in Portuguese)

Viitanen H (1996) Factors affecting the development of mould and brown rot decay in wooden material and wooden structures: effect of humidity, temperature and exposure time. Dissertation, Uppsala

Watt D, Colston B (2000) Investigating the effects of humidity and salt crystallization on medieval masonry. Build Environ 35(8):737–749

# Chapter 2
# Moisture Transfer Kinetics in Building Materials and Components: Modeling, Experimental Data, Simulation

**Antonia Moropoulou, Maria Karoglou, Asterios Bakolas, Magdalyni Krokida and Zacharias B. Maroulis**

**Abstract** Moisture is one of the most damaging factors of buildings. Masonries moisture content is defined by a delicate equilibrium established between the building and the environment. First order mathematical models can describe satisfactory phenomena like capillary rise and drying kinetics. Their advantages lie on the fact that they are general and include parameters with physical meaning. These parameters quantify the environment effect (temperature, relative humidity, air velocity) and the materials properties (porosity, density, average pore radius etc.). Constants introduced, such as capillary and drying time constant, correspond to the time that phenomena, capillary rise and drying, reach their equilibrium state. Furthermore, the same models can be used for the simulation of these phenomena on building components, leading to the development of a decision making tool for the design and selection of the most appropriate repair materials for masonries suffering by rising damp, based on materials microstructural properties.

**Keywords** Moisture • Building materials and components • Kinetic models • Capillary rise • Drying

---

A. Moropoulou · M. Karoglou (✉) · A. Bakolas · M. Krokida · Z. B. Maroulis
School of Chemical Engineering, National Technical University of Athens,
Zografou Campus, 15780 Athens, Greece
e-mail: margo@central.ntua.gr

A. Moropoulou
e-mail: amoropul@central.ntua.gr

A. Bakolas
e-mail: abakolas@mail.ntua.gr

M. Krokida
e-mail: mkrok@central.ntua.gr

Z. B. Maroulis
e-mail: maroulis@mail.ntua.gr

J. M. P. Q. Delgado (ed.), *Drying and Wetting of Building Materials and Components*,
Building Pathology and Rehabilitation 4, DOI: 10.1007/978-3-319-04531-3_2,

## Nomenclature

| | |
|---|---|
| $a_w$ | Air water activity (–) |
| $b_i$ | Oswin equations constants |
| $c_i$ | Drying kinetic constants |
| $D_w$ | Masonry width (m) |
| $D_p$ | Plaster width (m) |
| $\varepsilon_{cap}$ | Capillary active porosity (%) |
| $\varepsilon_o$ | Total porosity (%) |
| $\varepsilon_w$ | Wall total porosity (%) |
| $\varepsilon_p$ | Plaster total porosity (%) |
| $g$ | Gravitational constant (m/s) |
| $H$ | Capillary rise height (m) |
| $He$ | Capillary moisture equilibrium height (m) |
| $H_{ew}$ | Wall moisture equilibrium height (m) |
| $H_{ep}$ | Plaster moisture equilibrium height (m) |
| $H_{oi}$ | Capillary rise constants (m) |
| $m_{dry}$ | Dry mass of sample (g) |
| $m_c$ | Mass of sample saturated with water after capillary rise (g) |
| $m_i$ | Mass of sample saturated with water after immersion (g) |
| $n$ | Viscosity of water (Kg/sm) |
| $\rho_{bw}$ | Wall bulk density (kg/m$^3$) |
| $k_{cw}$ | Constant (kg/m$^2$d) |
| $k_{cp}$ | Constant (kg/m$^2$d) |
| $k_{dp}$ | Constant (kg/m$^2$d) |
| $\rho_{bp}$ | Plaster bulk density (kg/m$^3$) |
| $\rho_{sw}$ | Wall true density (kg/m$^3$) |
| $\rho_{sp}$ | Plaster true density (kg/m$^3$) |
| $\rho$ | Water density (g/cm$^3$) |
| $r_p$ | Plaster average pore radius ($\mu$m) |
| $r_w$ | Wall average pore radius ($\mu$m) |
| $s$ | Specific surface (m$^2$/g) |
| $\sigma$ | Water surface tension (dyn/cm) |
| $T$ | Air Temperature (°C) |
| $t_{cw}$ | Wall capillary time constant (d) |
| $t_{cp}$ | Plaster capillary time constant (d) |
| $t_c$ | Capillary time constant (d) |
| $t_d$ | Drying time constant (h) |
| $t_{dp}$ | Plaster drying time constant (h) |
| $u$ | Air velocity (m/s) |
| $V_{cum}$ | Total cumulative volume (mm$^3$/g) |
| $V_p$ | Total pore volume (mm$^3$) |
| $V_s$ | Material total volume (mm$^3$) |

$X_c$ Capillary moisture saturation content (kg/kg db)
$X_{cp}$ Plaster capillary equilibrium moisture content (kg/kg db)
$X_{cw}$ Wall capillary equilibrium moisture content (kg/kg db)
$X_{ep}$ Plaster equilibrium material moisture content
$X_i$ Total moisture saturation content (kg/kg db)
$X_o$ Initial moisture content (kg/kg db)

## 2.1 Introduction

Every year damages in buildings caused as direct or indirect effect due to the presence of moisture are estimated in millions of euro. Concerning historical buildings this problem is more intense, since they undergo long periods of wetting and drying cycles (Kieß 1989). In general moisture effect is related to materials durability and servile lifetime, energy saving in building, and human health (International Energy Agency 1990; Oliver 1997).

Any building is a complex structure consisting of various materials (stones, bricks, wood, metals, mortars, plasters, renders, coatings). Building materials can contain an amount of chemical bound water, for porous materials this percentage is 3–5 %, without affecting their durability. But if materials moisture content is over this percentage, then the deterioration effect of moisture is activated and can be of physical, chemical and biological nature (Oliver 1997; Avoletti 1997):

- Physical effect. In this case, major decay phenomenon is the formation of ice for low temperatures.
- Chemical effect. This is basically related to the salts transfer and crystallization. Salts are originated from the ground or the atmosphere. Some salts are hygroscopic, so they facilitate further water vapor absorption.
- Biological effect. Development of bio-deterioration, since most of microorganisms needs the presence of water in order to live and develop.

Main decay mechanisms are: hydrolysis, dissolution, hydration, oxidation, capillary rise, salt transfer and crystallization, hygroscopicity, cycles of wetting/drying; while main types of decay caused are: spalling, peeling, delamination, blistering, shrinkage, cracking, crazing, irreversible expansion, embrittlement, strength loss, staining discoloration, bio-decay of building materials (Connoly 1993).

Potential sources of water are: the ground, the environment (rain, sea, water vapor etc.), possible water sewage leakages, use of water for the production of building materials, interventions with the use of extensive quantities of water, hygroscopicity of salts (Oxley and Gobert 1998). Best way to "fight" moisture related problems in buildings is the prevention of the entrance of moisture at the design phase. However, the elimination of the problem in existing structures, especially historical ones, is more complicated. The various "anti-moisture" treatments used, can be either destructive or non destructive and their function is based mainly on two principles (Biscontin 1998; Massari and Massari 1981; Torraca 1981):

- To set a barrier at moisture entrance [like horizontal water-proofing with the use of water repellents or cutting of the masonry and use of a barrier of impermeable nature, creation of openings at the masonry (Koch technique) etc.]
- To increase the moisture evaporation rate (like air ventilation, knapen tubes, plasters that facilitate water evaporation, etc.)

For both principles moisture transport phenomena investigation is inevitable.

## 2.2 Water Transfer Mechanisms

### *2.2.1 Moisture Transfer Kinetics*

"Everything flows", used to say Greek philosopher Irakleitos. Kinetic models try to define flow position in space using time as a variable.

Water circulates in two phases, as vapor or liquid. The major moisture transfer mechanisms are at liquid phase: suction, diffusion, osmosis, electro kinesis, heat differences and at vapor phase: condensation, diffusion, hygroscopicity, evaporation and sorption (Torraca 1981). Masonry moisture content results by the dynamic equilibrium established between the masonry and the environment, as determined by the water transfer from the inner part of the masonry to the environment and vice versa. Very important is the fact that water can reach masonry in liquid or vapor form, but the only way to escape is in vapor form.

Deterministic modeling of this delicate equilibrium is very difficult, since the interaction between moisture and microstructure is very complicated, and the various moisture transfer mechanisms take place simultaneously, so the phenomena cannot be separated easily. In the literature various computerized hygro-thermal models are found that try to characterize, to describe changes in heat, air and moisture regimes within the wall assemblies (Canada Mortgage and Housing Corporation 2003). These models differentiate in the dimension and type of flow, the boundary conditions, building/components geometrical features, their commercially availability, their user-friendliness, the use of accurate material property data.

An appropriate approach can be the description of moisture transport phenomena, with the aid of phenomenological models, which include parameters of physical meaning. First order kinetic model can describe with satisfactory accuracy moisture transport of porous materials (Saravacos and Maroulis 2001). The simplicity of such models, including a small number of parameters, give them a general character applicable in many porous materials/components.

### 2.2.2 *Problem Definition*

Rising damp phenomenon is frequently encountered in buildings, especially historical ones. The mechanism of capillary rise phenomenon is since water carries soluble salts at the masonry. The maximum salts concentration, and thus the maximum deterioration, is observed at a specific zone of the masonry, which depends on the type of the building material and on the environmental conditions (Arnold 1982). Thus the capillary rise and drying kinetics is indicative of the susceptibility of materials to decay factors and can provide important information for the selection of compatible building materials for masonries. For the definition of moisture content and corresponding height of masonries suffering by rising damp are needed data concerning:

- Materials properties (micro-structural parameters, such as porosity, pore size, shape and distribution, water capacity currying)
- Experimental data and mathematical modeling of moisture transfer kinetics (capillary rise and drying kinetics)

## 2.3 Moisture Transfer Kinetics Modeling

### 2.3.1 *Materials Characteristics-Moisture Transfer*

Materials microstructure is a determinative factor concerning moisture transfer kinetics. In the literature are found many models that try to correlate pore structure with moisture transport coefficients. Most of these models are based at empirical correction factors, which are validated for a specific group of materials. Main difficulty derives by the fact that various pore sizes are defined by different techniques. Also the definition of pore size causes further difficulty. In addition the various moisture transfer mechanisms cannot be distinguished with accuracy, since transfer mechanism corresponds at pore range that in some times interpolate (Meng 1994).

One of the main parameters of microstructure is the total porosity of a material $\varepsilon_o$, equal to the ratio of the material empty space $V_b$ versus the material total volume $V_s$ (Amoroso and Camaiti 1997).

$$\varepsilon_o = 1 - \frac{V_b}{V_s} = 1 - \frac{\rho_b}{\rho_S} \tag{2.1}$$

where $\rho_b$, $\rho_s$ are the bulk and the true density of the material.

Pore structure, meaning pore shape, size, distribution, and networking of pores are very difficult to be defined. A very useful classification of pores to macro, meso and micro based on their diameter is shown in Table 2.1.

**Table 2.1** Pore classification based on their diameter *d* (nm) (IUPAC)

| Category | *d* (nm) |
|---|---|
| Macropores | ≥50 |
| Mesopores | 2–50 |
| Micropores | ≤2 |

There are different methods for the estimation of the porosity and miscrostructure parameters, like mercury intrusion porosimetry, helium pycnometry, digital image processing of microscopical images, x-ray and gamma ray attenuation (Amoroso and Camaiti 1997; International Union of Pure and Applied Chemistry 1994; Hall and Hoff 2002). Unfortunately there is not a technique that can measure the whole pore range. The combination of the results of each technique is also difficulty; due to the fact each technique uses different principles, so their result cannot be easily compared. Moreover if when they measure the same pore range, they measure different microstructure parameters.

### 2.3.2 Capillary Rise Height Kinetics Mathematical Modeling

Several studies attempt to optimize the existing theoretical models of capillary rise height kinetics. Most of the models attempt to model capillary moisture variations (Hall and Hoff 2002; Künzel 1995). Concerning the capillary rise height kinetics one of the simplest theoretical model, which relates the height penetration $H$ of a liquid into a capillary with pore radius $r$ to time, is the Lucas-Washburn equation:

$$H^2 = \frac{\sigma \cdot \cos\vartheta \cdot r}{2n} t \tag{2.2}$$

where, σ is the water surface tension, φ the contact angle of liquid–solid and n the fluid viscosity (Washburn 1921).

In this work a simple first order kinetic model is proposed to predict the capillary rise height $H$ as (Karoglou et al. 2005):

$$\frac{dH}{dt} = \frac{1}{t_c}(H_e - H) \tag{2.3}$$

where, $H_e$ is the equilibrium height and $t_c$ the capillary time height constant and $H_e$ the moisture equilibrium height.

The term of capillary height time constant $t_c$ is introduced for the first time and represents the time required for the waterfront of a material to reach 2/3 of the capillary moisture equilibrium height ($H_e$). The solution of Eq. (2.3) provides the development of height related to time:

**Table 2.2** Mathematical models for capillary rise and drying kinetics

| Capillary rise kinetic model | Drying kinetic model |
|---|---|
| *Properties* | |
| $H$ capillary rise height (cm) | X moisture content |
| $t_c$ capillary rise time constant (d) | $t_d$ drying time constant |
| *Factors* | |
| $H_e$ equilibrium vertical capillary rise height (cm) | $X_e$ equilibrium moisture content |
| $T$ air temperature (°C) | $T$ air temperature (°C) |
| $r$ average pore radius (μm) | $\alpha_w$ air water activity (–) |
| | $u$ air velocity (m/s) |
| *Properties equations* | |
| $\frac{dH}{dt} = \frac{1}{t_c}(H_e - H)$ | $\frac{dX}{dt} = -\frac{1}{t_d}(X - X_e)$ |
| $H = H_e - (H_e - H_o)\exp\left(-\frac{t}{t_c}\right)$ | $t_d = c_o\, T^{c_1}\, \alpha_w^{c_2}\, u^{c_3}$ |
| | $t_d = \frac{1}{K}$ |
| | $t = t_d \ln\left(\frac{X-X_e}{X_o-X_e}\right)$ |
| $H_e = \frac{2\sigma}{\rho g r}$ | $X_e = b_o \exp\left(\frac{b_1}{T}\right)\left(\frac{a_w}{1-a_w}\right)^{b_2}$ |
| *Parameters* | |
| $H_o$ correction factor (cm) | *Constants* $c_0$(h), $c_1$(–), $c_2$(–), $c_3$(–) |
| $\rho$ water density (g/cm$^3$) | *Constants* $b_0$ (kg/kg db %), $b_1$ (°C), $b_2$ (–) |
| $\sigma$ water surface tension (dyn/cm) | |
| *Factors affecting parameters* | |
| Material → $H_o$ | Material → $c_0$, $b_0$ |
| Environment → $\rho$, $\sigma$ | Environment → $c_1$, $c_2$, $c_3$, $b_1$, $b_2$ |

$$H = H_e - (H_e - H_o)e^{-\frac{t}{t_c}} \tag{2.4}$$

where, $H_o$ is a correction factor.

For the estimation of moisture equilibrium height $H_e$ for vertical rise it is used Jurin law, corresponding to the point that capillary suction pressure equals to hydraulic pressure (Gennes et al. 2004).

$$H_e = \frac{2\sigma}{\rho g r} \tag{2.5}$$

where, $r$ the mean radius of the capillary, $\sigma$ the surface tension of the liquid, $\rho$ water density and $g$ gravity acceleration force.

Therefore, if the material pore radius is known, the theoretical capillary moisture equilibrium height can be easily estimated. It should be noted that the pores of the materials are assumed to have a cylindrical shape and present homogeneous microstructure. Furthermore, evaporation effects are disregarded and only vertical rise of water is considered. The mathematical model of capillary rise kinetics is summarized in Table 2.2.

### 2.3.3 Drying Kinetics Mathematical Modeling

Many moisture transfer models have been used to describe the drying process of porous materials (Zaknoune et al. 2012; Dantas et al. 2003; Dietl et al. 1998). A first order drying kinetic model is appropriate to describe mass transfer phenomena. The drying time constant depends on both material and drying air properties, as it is the phenomenological property representative of several transport phenomena. The effect of air temperature, relative humidity and air velocity on the drying time constant of building materials has been studied partially (Karoglou et al. 2005). The drying rate of a product depends on the material characteristics and the drying air conditions. The drying rate may indicate initially a constant rate period, during which free water is removed.

Independently of the physical drying mechanism, in most cases a first order kinetic model is adequate to describe the falling rate drying period:

$$\frac{dX}{dt} = -\frac{1}{t_d}(X - X_e) \tag{2.6}$$

where, $X$ is the material moisture content at time $t$, $t_d$ is the drying time constant and $X_e$ is the equilibrium material moisture content (Saravacos and Maroulis 2001).

The drying time constant is consistent with the process control terminology for first-order dynamic systems and its value has physical meaning denoting the required time to remove the 2/3 of the total moisture. The drying time constant $t_c$ depends on drying air conditions and material characteristics. A simple power law model seems adequate for most purposes. It incorporates the effect of drying air conditions that is air velocity $u$, temperature $T$ and water activity $a_w$ (Marinos-Kouris and Maroulis 1995):

$$t_d = c_o T^{c_1} \alpha_w^{c_2} u^{c_3} \tag{2.7}$$

where, $c_0$, $c_1$, $c_2$, $c_3$ are adjustable empirical constants, depending on the material characteristics and the drying equipment geometry.

The drying time constant $t_d$, is used instead of the well known drying constant $K$:

$$t_d = \frac{1}{K} \tag{2.8}$$

When drying conditions are kept constant, Eq. (2.6) is integrated as follows:

$$t = t_d \ln\left(\frac{X - X_e}{X_o - X_e}\right) \tag{2.9}$$

Equation (2.9) calculates the required drying time t to remove moisture from the initial $X_o$ to the final $X$ material moisture content.

The equilibrium moisture content $X_e$ is usually measured with a steady state approach, where the material in question is brought to equilibrium in an atmosphere of a known relative humidity and the weight is noted. This may be done through an adsorption and desorption cycle. The principal factors affecting hysteresis type and shape have been summarized by Brunauer, Emmet and Teller (BET) into 5 main categories and are related to materials microstructure. Some types possess a hysteresis loop, the lower branch of which represents measurements obtained by progressive addition of gas to the system, and the upper branch by progressive withdrawal (Mangel 1999). For porous solids, such as building materials, modified Oswin model is more suitable for estimation of equilibrium moisture content $X_e$ as (Karoglou et al. 2005):

$$X_e = b_o \exp\left(\frac{b_1}{T}\right)\left(\frac{a_w}{1-a_w}\right)^{b_2} \tag{2.10}$$

where, $T$ is the dry bulb temperature of air (°C), $a_w$ is the water activity of the air and $b_0$, $b_1$, $b_2$ are empirical constants.

The values of the constants $b_0$, $b_1$, $b_2$ can be determined by fitting the proposed equation to the experimental data, minimizing the standard deviation between predicted by the model and experimental values.

In Table 2.2 are summarized the mathematical models for drying kinetics.

## 2.4 Experimental Techniques, Data and Results

Two main categories of materials are examined; materials serving as masonry substrate materials, such as bricks/stones and plasters. In general bricks are symbolized with the starting letter *B*, stones with *S* and plasters with *P*. The bricks are of traditional type (*BRI*, *BRM*), while stones of quarries, from Crete (*SRY*, *SRW*), Rhodes (*SRH*) and Cyprus (*SCY*). Regarding plasters, three types are investigated. A traditional plaster taken from a neoclassical building at the center of Athens (*PZN*), common cement based plaster used in new constructions and restoration works in Greece (*PTI*) and four premixed plasters of controlled microstructure, suitable for masonries suffering from rising damp (*PEM*, *PMP*, *PRL*, *PTR*). A series of measurements are performed in order to investigate materials microstructural parameters and the estimation of capillary rise and drying kinetics parameters.

### *2.4.1 Microstructure Parameters Investigation*

For the study of materials microstructural characteristics mercury intrusion porosimetry is used. With the aid of this technique are estimated; total cumulative volume $Vcum$, specific surface $s$, average pore radius $r$, bulk density $\rho_{bulk}$, corrective bulk density $\rho_{bulkcor}$ and total porosity $\varepsilon_o$ (%).

Materials water capacity caring, as described by the total saturation moisture $X_i$ (%) content depends, is related to materials porosity and is estimated by the gravimetric method (g/g dry base) as:

$$X_i\ (\%) = \left(\frac{m_i - m_{dry}}{m_{dry}}\right)(\%) \tag{2.11}$$

where, $m_i$ is the weight of sample after the immersion in water and $m_{dry}$ the initial dry mass of the sample.

Moreover the capillary saturation moisture content is the amount of water obtained by capillaries until the water reached the upper surface of the sample. The capillaries are the pores, which play the most significant role in the capillary rise kinetics. If a material has a high percentage of capillaries, it is obvious that the capillary rise rate will be also high. At the end of a capillary rise experiment when the sample is saturated the capillary moisture content, $Xc$ (%), can be calculated in dry base as:

$$X_c\ (\%) = \left(\frac{m_c - m_{dry}}{m_{dry}}\right)(\%) \tag{2.12}$$

Consequently, this percentage could explain the reason for the differences in the capillary rise kinetics for different materials. A way to estimate this percentage is to calculate the capillary active porosity $\varepsilon_{cap}$, which is a factor that can be indicative for the capillaries:

$$\varepsilon_{cap} = \frac{X_c\ (\%)}{X_i\ (\%)}\varepsilon_o \tag{2.13}$$

where, $Xc$ the capillary saturation moisture content (g/g dry base) obtained by capillary rise, $Xi$ the amount of water obtained by total immersion (g/g dry base) and $\varepsilon_o$ (%) is the porosity of the material.

In the following table mercury porisimetry results are shown (total cumulative volume $V_{cum}$, specific surface $s$, average pore radius $r$, bulk density $\rho_{bulk}$, corrective bulk density $\rho_{bulkcor}$ and total porosity $\varepsilon_o$ (%) and capillary active porosity $\varepsilon_{cap}$ (%) for the 12 materials under investigation (Table 2.3).

Regarding bricks, *BRI* presents high porosity value and low average pore radius and pores distributed around average pore radius, while *BRM* presented lower porosity values than *BRI*, but higher average pore value and a wider distribution of pores as presented in Fig. 2.1. Moreover for *BRI* the values of $\varepsilon_o$ and $\varepsilon_{cap}$ differ 1.2 %, showing that most of its pores are capillaries. For the stones *SRW* presents the highest porosity values ~35 (%). Smallest difference between total porosity and active capillary porosity presented the *SRH* equal to 3.4, which stone also presented the highest average pore radius. Regarding plasters smallest difference between total porosity and active capillary porosity presented cement plaster *PTI* equal to 2.8. *PEM* presented the highest values of porosity and average pore radius. *PTR* although presented high porosity values it presented very low values of capillary active porosity, because it contained hydrophobic agents.

**Table 2.3** Mercury intrusion parameters and capillary active porosity results

| Code | $V_{cum}$ (mm$^3$/g) | $s$ (m$^2$/g) | $r$ (μm) | $\rho_{bulk}$ (g/cm$^3$) | $\rho_{bulkcor}$ (g/cm$^3$) | $\varepsilon_o$ (%) | $\varepsilon_{cap}$ (%) |
|---|---|---|---|---|---|---|---|
| *BRI* | 323 | 4.88 | 0.45 | 1.53 | 3.01 | 49.2 | 48.0 |
| *BRM* | 174 | 1.46 | 2.01 | 1.87 | 2.77 | 32.6 | 27.0 |
| *SCY* | 136 | 2.75 | 4.86 | 2.05 | 2.80 | 27.7 | 19.0 |
| *SRH* | 90.2 | 1.34 | 39.8 | 2.45 | 3.14 | 22.0 | 18.6 |
| *SRY* | 197 | 2.28 | 2.83 | 1.79 | 2.77 | 35.2 | 30.8 |
| *SRW* | 180 | 1.67 | 4.45 | 1.87 | 2.81 | 33.7 | 29.8 |
| *PEM* | 408 | 6.97 | 33.2 | 1.40 | 4.40 | 63.8 | 38.2 |
| *PMP* | 161 | 6.36 | 2.42 | 1.96 | 2.89 | 32.4 | 25.1 |
| *PRL* | 199 | 5.15 | 0.87 | 1.75 | 2.69 | 34.8 | 31.6 |
| *PTI* | 273 | 3.19 | 1.46 | 1.54 | 2.66 | 42.1 | 39.3 |
| *PTR* | 712 | 15.0 | 0.53 | 0.89 | 2.40 | 63.1 | 10.0 |
| *PZN* | 213 | 1.76 | 0.52 | 1.94 | 3.25 | 41.3 | 36.9 |

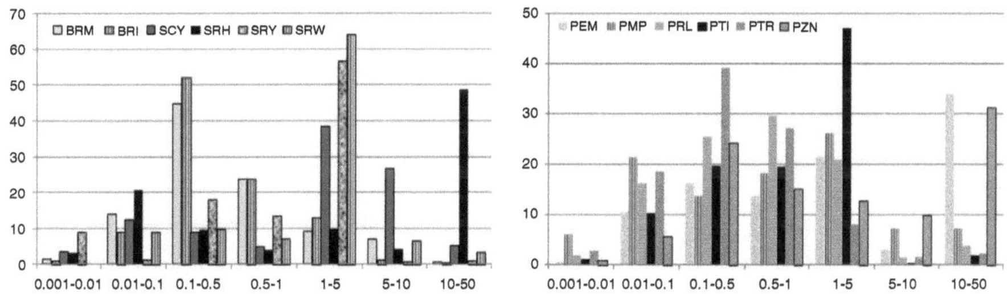


**Fig. 2.1** Distribution of the relative pore volume (%) at various pore radius ranges in the *left* for bricks and stones and in the *right* for plasters

## 2.4.2 Capillary Rise Height Kinetic Results

There is no standard describing an experimental procedure for obtaining the capillary rise height related to time. The capillary height can be measured directly when performing a typical capillary rise test by measuring changes of height in time, or if measuring moisture content, the corresponding height can be extracted by the following equation, considering that moisture content and height are proportional magnitudes:

$$H = \frac{X}{X_c} H_{sample} \tag{2.14}$$

where, $X$ is the moisture content in time $t$ (g/g dry base), $X_c$ the capillary saturation moisture content (g/g dry base) and $H_{sample}$ the geometric height of the sample.

The first order kinetic model (Table 2.2) is fitted to experimental data and the results of parameter estimation are summarized in Table 2.4. The parameters of the proposed model ($H_o$, $t_c$) result from an optimization technique to minimize the

**Table 2.4** Capillary rise estimated parameters

| Code | $H_e$ (cm) | $H_o$ (cm) | $t_c$ (d) |
|---|---|---|---|
| *BRI* | 3.333 | 0.09 | 3.81 |
| *BRM* | 667 | 0.06 | 1.88 |
| *SCY* | 309 | 0.14 | 0.79 |
| *SRH* | 38 | 0.32 | 0.02 |
| *SRY* | 564 | 0.00 | 1.27 |
| *SRW* | 320 | 0.16 | 0.24 |
| *PEM* | 49 | 0.17 | 0.04 |
| *PMP* | 649 | 0.22 | 0.70 |
| *PRL* | 1.724 | 0.07 | 2.10 |
| *PTI* | 1.030 | 0.05 | 1.97 |
| *PTR* | 3.000 | 0.68 | 2.27 |
| *PZN* | 2.941 | 0.06 | 5.93 |

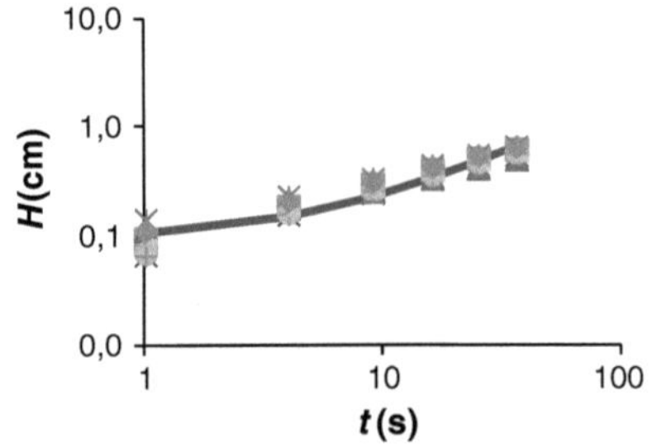


**Fig. 2.2** Fitting of capillary rise model to experimental data for the estimation of capillary rise time constant $t_c$ at the first 100 s for brick *BRI*

residual sum of squares between experimental and calculated values of capillary height. The capillary time constant is calculated for experimental times until 100 s. Each set of parameters corresponds to a different building material.

In Fig. 2.2 is shown the experimental and theoretical values for capillary time constant $t_c$ for brick *BRI*.

In the Fig. 2.3 is shown the theoretical changes of capillary rise height. In plasters category *PEM* and *PMP* reach faster their theoretical moisture equilibrium height, while concerning substrate materials, *SRH* and *SRW*.

### 2.4.3 Drying Kinetics: Sorption Isotherms

Experiments to determine the influence of process variables on the drying kinetics can be performed in a convective air dryer. The variables taken into consideration are air temperature, humidity and velocity. The drying experiments are carried out, at four levels of air-temperature (20°, 30° and 35 and 40 °C), three levels of air velocity (1, 3 and 8 m/s) and 5 levels of relative humidity of the air (30–80 %).

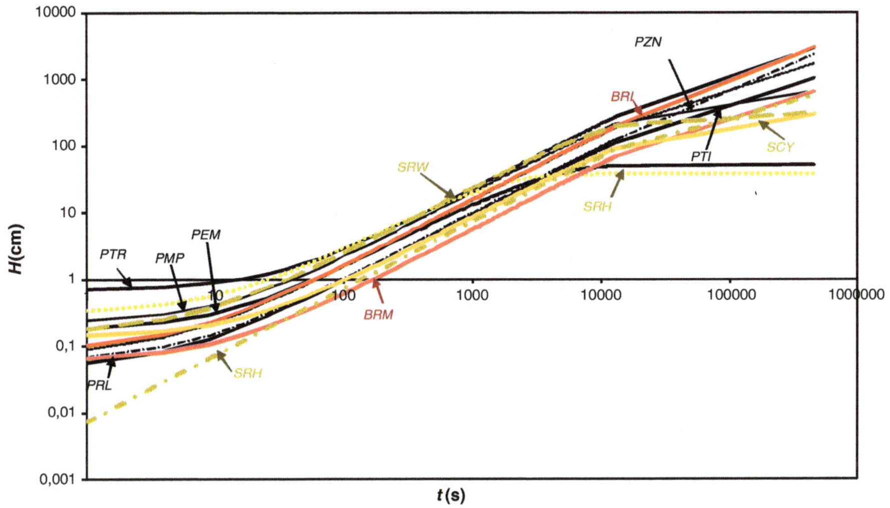


**Fig. 2.3** Theoretical height versus time for the various materials

Before starting the experiment, samples are saturated through immersion in distilled water and the $X_i$ (%) is measured for each material. The model parameters are estimated by minimizing the standard deviation between the proposed mathematical model and the experimental data. The results of fitting the proposed mathematical model to the experimental data are summarized in Table 2.5. For $c_1$ parameter which represents the temperature effect at drying time constant is given the value 0, because the temperature effect is included at the estimation of moisture equilibrium content $X_e$ at the modified Oswin model. Parameter $c_3$, which is correlated with the effect of air velocity based on previous research results, has a value of −0.8.

A paradigm of the results of drying kinetics is shown in next figures showing the fitting of the drying kinetics model to the experimental data for variations of relative humidity and air velocity for the plaster *PRL* (Fig. 2.4).

For all materials the effect of air relative humidity and air velocity is the same, when relative humidity increases the drying rate decreases, while when the air velocity increases the drying rate increases too.

For calculating sorption parameters $b_0$, $b_1$, and $b_2$ parameters moisture sorption, adsorption and desorption isotherms, are determined using a water sorption analyzer, at three different temperatures (15, 25 and 35 °C). The hysteresis phenomenon between adsorption and desorption, is observed for all the examined materials. The modified Oswin equation is fitted to the experimental data for water activity ranging between 0.03–0.90. Statistical analysis shows that the effect of the environment was constant for all the examined materials. In Fig. 2.5 is shown the experimental and the predicted values of water absorption and desorption for the plaster *PRL*.

**Table 2.5** Estimated parameters for drying and sorption parameters

| Code | $c_0$ (h) | $c_2$ (–) | $b_0$ (kg/kg db %) | $b_1$ (–) | $b_2$ (h) |
|---|---|---|---|---|---|
| *BRI* | 1.92 | 0.89 | 3.53E–02 | 0.57 | 0.51 |
| *BRM* | 1.09 | 0.89 | 1.46E–05 | 2.33 | 0.71 |
| *SCY* | 0.87 | 0.47 | 2.39E–02 | 1.16 | 0.30 |
| *SRH* | 0.78 | 0.87 | 3.19E–05 | 2.49 | 0.45 |
| *SRY* | 1.16 | 0.85 | 3.10E–02 | 0.73 | 0.41 |
| *SRW* | 0.87 | 0.72 | 3.53E–05 | 0.57 | 0.51 |
| *PEM* | 1.30 | 1.01 | 1.44E–04 | 2.55 | 0.38 |
| *PMP* | 0.91 | 0.84 | 2.00E–06 | 3.60 | 0.44 |
| *PRL* | 1.17 | 0.85 | 1.69E–02 | 1.10 | 0.44 |
| *PTI* | 1.33 | 0.78 | 5.40E–03 | 1.31 | 0.35 |
| *PTR* | 1.73 | 0.83 | 7.17E–06 | 3.85 | 0.31 |
| *PZN* | 1.15 | 0.67 | 8.61E–07 | 3.79 | 1.03 |

$c_1 = 0(–), c_3 = -0.8(–)$

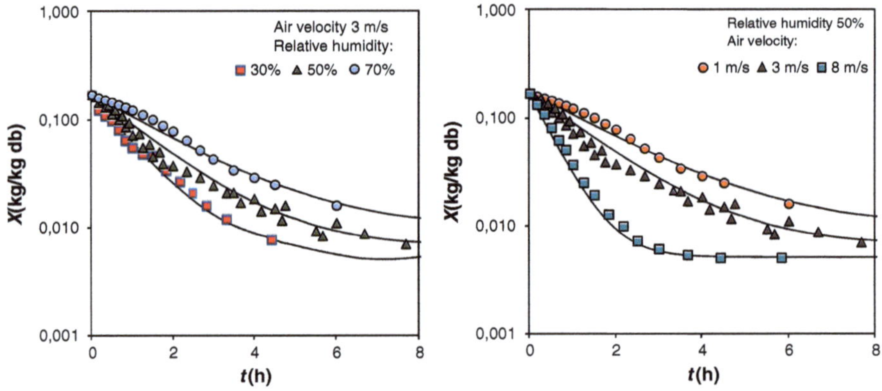


**Fig. 2.4** Experimental data and predicted values for variations of air relative humidity and air velocity for plaster *PRL*

Practically adsorption–desorption data take place simultaneously. Mean adsorption–desorption data called as sorption data can be described by a mean curve. A second fitting to experimental data is made, for the sorption data. The results estimated for parameters $b_0$, $b_1$, $b_2$ for sorption curves are shown in Table 2.5.

In the following figure is shown the variation of drying time constant as a function of relative humidity (with constant air velocity) for all materials under investigation. Higher values presented for plasters *PEM* and *PMP*, while for substrate material the stone *SRH*. These materials presented also the higher average pore radius values (Fig. 2.6).

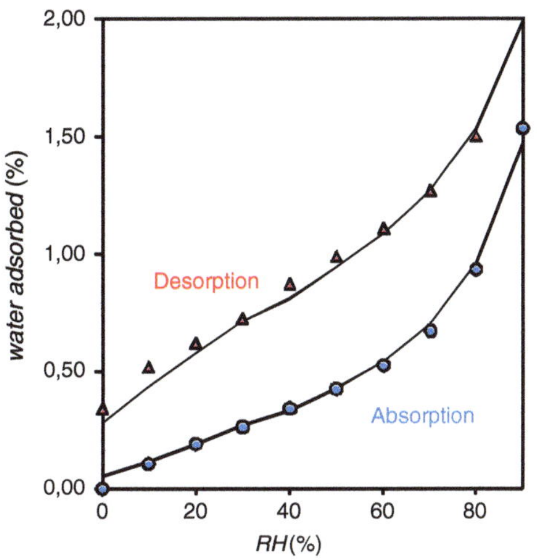


Fig. 2.5 Experimental data and predicted values for absorption and desorption isotherms of plaster PRL at 25 °C

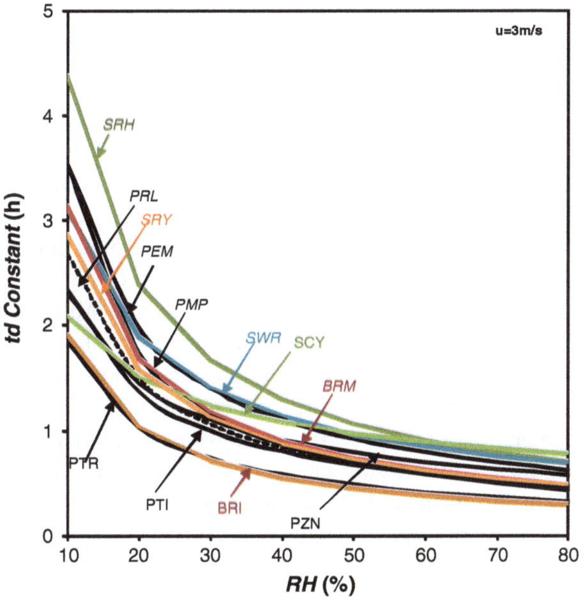


Fig. 2.6 Effect of relative humidity on the drying time constant for all building materials for constant air velocity

## 2.5 Simulation

### *2.5.1 Simulation Mathematical Modeling*

A simulator is developed that takes into account the moisture transfer mechanisms to and from the building (capillary rise, drying), the wall configuration (materials and size), the construction materials properties (porosity, density etc.) (d) the seasonal

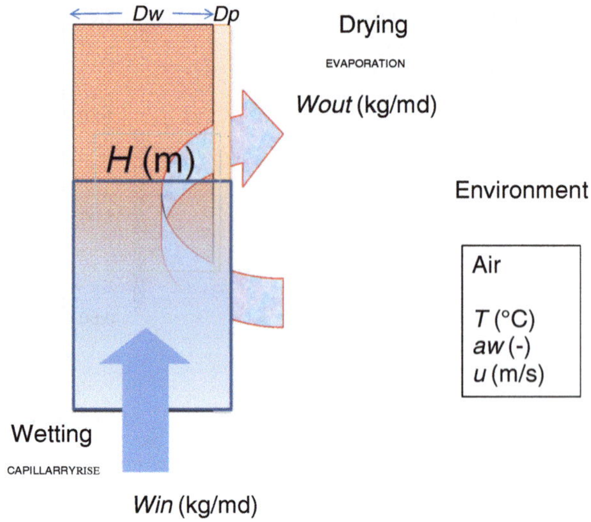


**Fig. 2.7** Masonry system geometry and moisture transfer main parameters

region meteorological data (air temperature, humidity and velocity); and calculates: (a) the seasonal wall moisture content along with the corresponded equilibrium moisture height; (b) the capillary rising water flow rate; (c) the wall drying flow rate (Konstantopoulos et al. 2007). The simulator has been developed for a masonry that consists of two layers, the substrate wall and the outer plaster. Wall and plaster are compact and homogeneous, consisting of one material. Wall is constructed either with brick or stone. The above masonry system is in contact with ground moisture, while only the plaster layer is in contact with air. In Fig. 2.7 is shown the main mechanisms and characteristics of the two layer masonry. Wall width is $D_w$, and plaster width $D_p$. The wall is constructed of either stone or brick and the outer layer of a plaster with various compositions. Materials present homogenous microstructure.

The main moisture transfer mechanisms taken into consideration are the capillary rise and the drying of ground moisture. Capillary rise is described by flow rate, $W_{in}$ (kg/md), while drying rate is described by $W_{out}$ (kg/md). Capillary moisture height $H$ (m) is considered the same for both layers. The system is in equilibrium with the environment, and the equilibrium moisture storage depends on the environmental conditions: air temperature $T$ (°C), water activity, $\alpha_w$ (–) and air velocity, $u$ (m/s).

The mathematical model of the developed simulator is shown in Table 2.6.

### *2.5.2 Simulator Architecture*

The Simulator is an application developed under Visual Studio 2005 using the C# programming language. The simulator is a combination of three basic components:

- Database
- Mathematical Model Library
- Graphical User Interface (GUI)

**Table 2.6** Simulator mathematical model

| Simulator mathematical model | | | |
|---|---|---|---|
| *Materials properties* | | | |
| $\rho_{bw} = (1 - \varepsilon_w)\rho_{sw}$ | (2.15) | $\rho_{bp} = (1 - \varepsilon_p)\rho_{sp}$ | (2.16) |
| *Wall* | | *Plaster* | |
| $\varepsilon_w$ (%) total porosity | | $\varepsilon_p$ (%) total porosity | |
| $\rho_{sw}$ (kg/m$^3$) true density | | $\rho_{sp}$ (kg/m$^3$) true density | |
| $r_w$ (μm) average pore radius | | $r_p$ (μm) average pore radius | |
| *Capillary rise kinetics* | | *Drying kinetics* | |
| $H_{ew} = \frac{2\sigma}{\rho g r_w}$ | (2.17) | $X_{ep} = b_o \exp\left(\frac{b_1}{T}\right)\left(\frac{a_w}{1-a_w}\right)^{b_2}$ | (2.18) |
| $H_{ep} = \frac{2\sigma}{\rho g r_p}$ | (2.19) | $t_{dp} = c_o T^{c_1} \alpha_w^{c_2} u^{c_3}$ | (2.20) |
| $X_{cw}$ (kg/kg db) wall equilibrium capillary moisture content | | | |
| $X_{cp}$ (kg/kg db) plaster equilibrium capillary moisture content | | | |
| $t_{cw}$ (d) wall capillary time constant | | | |
| *Masonry moisture balance* | | | |
| $H = \frac{k_{cw}H_{ew}+k_{cp}H_{ep}}{k_{cw}+k_{cp}+k_{dp}}$ | (2.21) | $k_{cw} = \frac{\rho_{bow}D_w X_{cw}}{t_{cw}}$ | (2.22) |
| | | $k_{cp} = \frac{\rho_{bop}D_p X_{cp}}{t_{cp}}$ | (2.23) |
| | | $k_{dp} = \frac{X_{cp} \quad X_{ep}}{t_{dp}} \rho_{bop} D_p$ | (2.24) |
| *Environmental conditions* | | | |
| $T$ (°C) air temperature | | | |
| $a_w$ (–) air water activity | | | |
| $u$ (m/s) air velocity | | | |
| *Empirical constants* | | | |
| $H_{ow}$, $H_{op}$ capillary rise kinetics equ ation | | | |
| $b_0$, $b_1$, $b_2$ Oswin equation for materials isotherms | | | |
| $c_0$, $c_1$, $c_2$, $c_3$ drying kinetics equation | | | |

The first component is the database, where material and environmental data are stored. The terms material and environmental data correspond to a collection of attributes that fully define a set of materials and a set of environmental conditions. For material data these attributes are: total porosity, density, capillary saturation moisture, equilibrium moisture constant, drying constant, average pore radius, capillary time constant and the name of the material. For environmental conditions data these attributes are: the minimum and maximum temperature, moisture and air velocity values that appear in the specified environment, the date when the maximum for each of the above values is observed, daily observed temperature, moisture and air velocity conditions.

The second component is the mathematical model library embedded in the application. This library is the core of the simulator since it brings into effect the methods and the principles described previously.

The third component of the application is the Graphical User Interface (GUI). The simulator is a Multiple Document Interface (MDI) application which means that the environment under which the user has to work has a desktop-like behavior.

The GUI consists of four basic windows (screens), which are presented in greater detail in the following section. The GUI enables the user to interact with the database in various ways (create, delete and update instances) and also create and manage simulation projects. At last the user can customize the application settings through the settings window.

## *2.5.3 Graphical User Interface*

### 2.5.3.1 Simulation Project Window

To start a new simulation project, the user has to select the File—New—Simulation Project menu. This starts a new blank simulation project and opens the relative window which will host the project (Fig. 2.8). In the bottom left (Fig. 2.8, part 3) area are located the controls with which the user can interact to manage the simulation project. With the aid of drop down menus the user can choose the masonry dimensions (wall and plaster width and masonry height), the material used as wall material and the plaster material, and also the place the masonry is.

After the desired values for these controls have been set, the simulator calculates the curve for the moisture height (by using the embedded mathematical library) and presents it to the graphic area located in the upper right part of the window (Fig. 2.8, part 2). This is the curve that depicts the height of the moisture (vertical axis) relative to time (horizontal axis) for all 365 days of the year. From this curve the user can also see the minimum and maximum heights of moisture during the whole year. On top of this graphic area are two tabs from which the user can switch between the curve view and the report view, where he can see the moisture height results of the calculations. In the lower right part of the window there is a graphic design of the wall with the plaster material attached. This graphic design is in scale and shows the geometrical aspects of the wall—plaster (Fig. 2.8, part 4). It also shows the minimum and maximum height of moisture during the year once the calculations have finished. In the upper left part of the window is another graphic area that shows the measurements curve and the proximal sinus curve of the selected environment (Fig. 2.8, part 1). The user can select one of the three tabs on top of this graphic area to change among the temperature, moisture and air velocity curves.

### 2.5.3.2 Material Manager Window

This window can be used to interact with the material data of the database. When the application starts up the default material library is loaded. The user can select a material (Fig. 2.9, part 1) and can see the current attribute values of the selected material (Fig. 2.9, part 3). The user can also select an attribute and change its

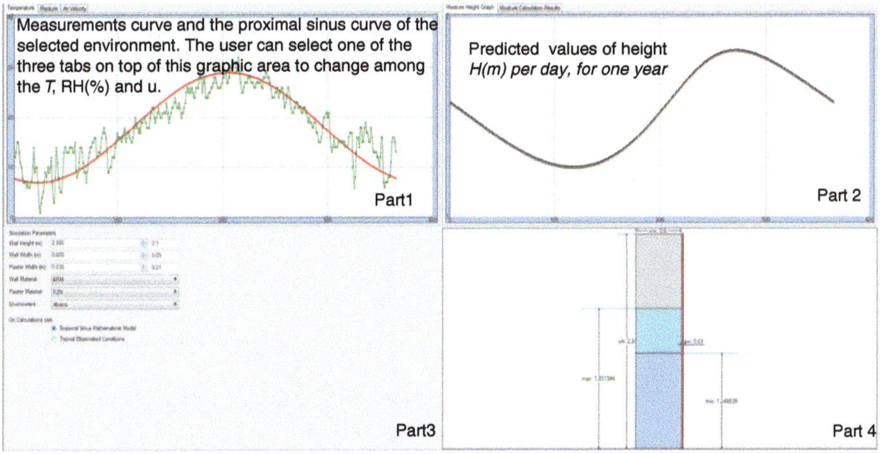


**Fig. 2.8** Simulation project window

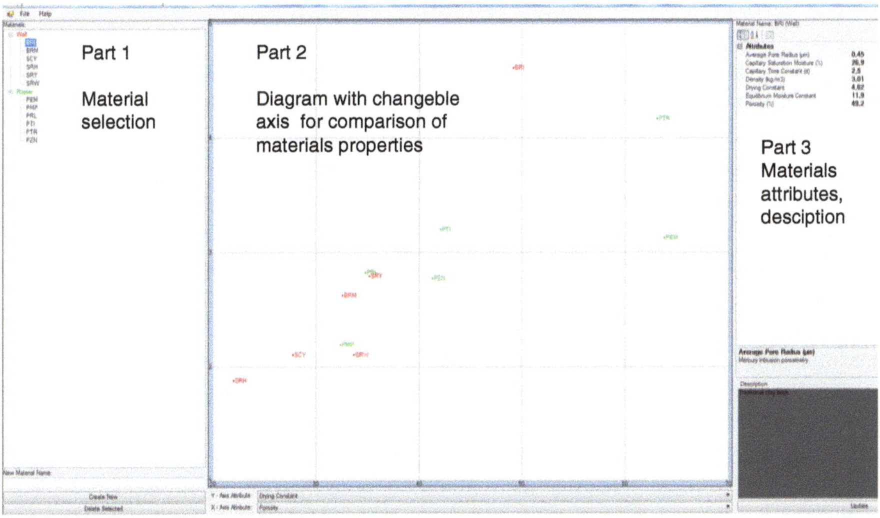


**Fig. 2.9** Material manager window

value to a desired one. Below the property list is a help area where the user can see a short description for the selected material attribute. In the center of the window is a graphic area where all materials are depicted in a two dimensional graph (Fig. 2.9, part 2). Each axis of the graph represents a specific material attribute which can be assigned from the two list controls located at the bottom of the window. This graph is very useful because the user can see how materials relate to one another according to the selected attributes that each axis represents.

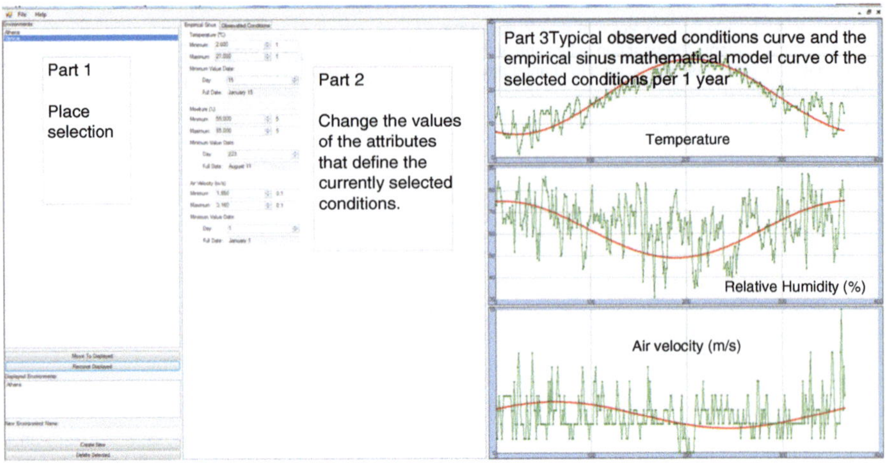


**Fig. 2.10** Environmental manager window

#### 2.5.3.3 Environmental Manager Window

Though this window, the user can interact with the environmental data of the database. In the left part of the window there is a list that contains the names of the sites (environmental conditions) in the loaded library (Fig. 2.10, part 1). Below the list there are controls through which the user can create or remove environmental conditions from the library. In the center of the window there are controls which the user can manipulate to change the values of the attributes that define the currently selected conditions (Fig. 2.10, part 2). Above these controls the user can choose between two tabs, the Empirical Sinus Mathematical Model tab and the Typical Observed Conditions tab. These tabs switch the controls that appear below between the ones that manipulate the proximal sinus parameters and an array control where the user can update the measurement values for the currently selected conditions. In the right part of the window there are three graphic areas (Fig. 2.10, part 3). These areas are used to represent visually the typical observed conditions curve and the empirical sinus mathematical model curve of the selected conditions. Note that there might be more than one environments selected. If this is the case, the graphic areas will represent multiple curves, a functionality that enables the user to compare the environmental conditions curves (both types) for several environments at the same time. The three graphic areas correspond to the three types of attributes that define a site: temperature, moisture and air velocity.

#### 2.5.3.4 Simulation Case Study

The simulator is a powerful tool for the evaluation and optimization of masonry design characteristics concerning rising damp phenomena in buildings. The proposed simulator supports satisfactory decision making concepts concerning the selection of the appropriate protecting strategy, e.g., plaster selection (material, size, replacing time).

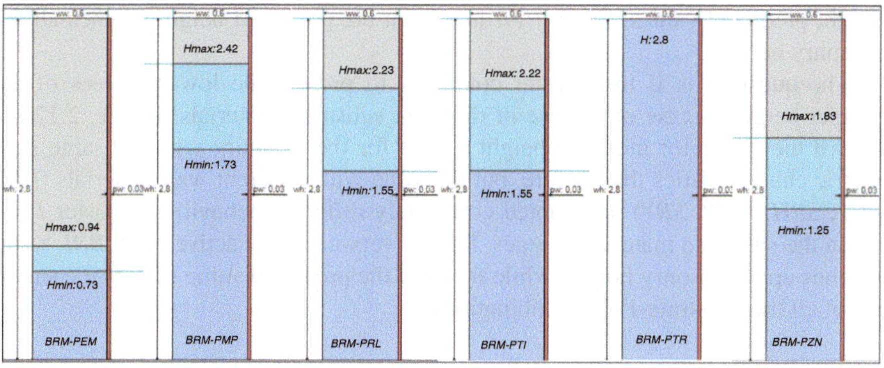


**Fig. 2.11** Minimum and maximum predicted capillary rise heights with different plasters

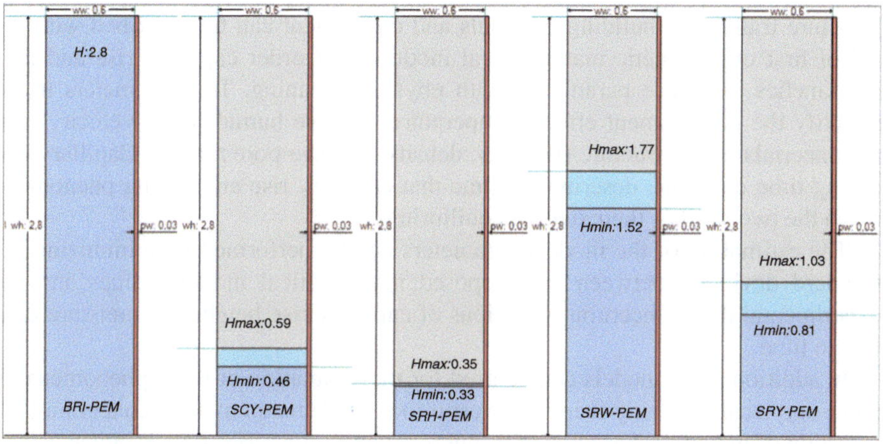


**Fig. 2.12** Minimum and maximum predicted capillary rise heights for masonry with *PEM* plaster using different substrate materials

With the use of this tool, the assessment of the effectiveness of restoration materials before their real application can be made, contributing to the extension of masonries lifetime. The appropriate restoration materials choice is one way to control the masonry moisture equilibrium. As an example, is considered masonry with the following characteristics: height 2.8 m, wall width equal to 0.6 m and plaster application width 0.03 m. The masonry is situated at Athens. Two different scenarios are investigated:

- When the substrate material is known, which plaster presents the better anti-moisture performance?
- How this plaster performs in case of different substrates?

In first case as substrate material is selected *BRM*. The predicted heights after the selection of the different plasters are shown in Fig. 2.11 and can be easily concluded that for *PEM* resulted the lowest minimum and maximum capillary height predicted values. When *PTR* is chosen the predicted height reaches all the masonry height.

The question is if this plaster continues to present the lowest values of the predict height in case of the use of different substrate materials. In Fig. 2.12 are shown the simulator moisture height results for the masonry with the same geometric characteristics than before, but this time with different wall materials (*BRI*, *SCY*, *SRH*, *SRW*, *SRY*). It is noted completely different behavior of plaster *PEM* when the substrate material changes. For the very capillary active brick *BRI*, water reaches up to masonry height, while for *SRH* the predicted value of *H* is the smallest of all the substrate-*PEM* combinations.

## 2.6 Conclusions

Moisture transfer in building materials and component can be described with the aid of first order kinetic mathematical models. Fist order capillary rise and drying kinetics introduce parameters with physical meaning. The parameters try to quantify the environment effect (temperature, relative humidity, air velocity) and the materials microstructure (porosity, density, average pore radius). Capillary and drying time constants describe the time that capillary rise and drying phenomena reach the two third of their state of equilibrium.

The estimation of the models parameters can be performed by minimizing the standard deviation between the proposed mathematical models values and the experimental data concerning variations of capillary rise height and moisture content to time.

In addition these models can be used for the simulation of these phenomena on a masonry scale, making them a powerful tool for decision making concerning the selection of materials, based on their microstructure for various environments.

## References

Amoroso GG, Camaiti M (1997) Scienza dei materiali e restauro. ALINEA, Bolgona

Arnold A (1982) Rising damp and saline minerals. In: 4th international conference on the deterioration and preservation of stone objects, Louisville, Kentucky, pp 11–28

Avoletti M (1997) Contro l'umitida' ascendente. Recupero & Conservazione 17:58–705

Biscontin G (1998) La risalita capillare dell'acqua nelle murature: note per la diagnosi e il risanamento. Recupero edilizio, Umidita Techniche e prodotti di risanamento, ALINE EDITRICE, pp 61–67

Canada Mortgage and Housing Corporation (2003) Review of hygro-thermal models for building envelope retrofit analysis, research highlights. Technical series 03-128. http://www.cmhc-schl.gc.ca/publications/en/rh-pr/tech/03-128-e.htm

Connoly JD (1993) Humidity and building materials. In: Rose WB, Tenwolde a (eds) Bugs, mold and rot II, proceedings of a workshop by the building environment and thermal envelope council of the National Institute of Building Sciences, Washington DC, NIBS, pp 29–36

Dantas LB, Orlande HRB, Cotta RM (2003) An inverse problem of parameter estimation for heat and mass transfer in capillary porous media. Int J Heat Mass Transfer 46:1587–1598

Dietl C, Winter E, Viskanta R (1998) An efficient simulation of heat and mass transfer processes during drying of capillary porous hygroscopic materials. Int J Heat Mass Transfer 41:3611–3625

Gennes PG, Brochard-Wyart F, Quere D (2004) Capillarity and wetting phenomena: drops, bubbles, pearls, waves. Springer, New York

Hall C, Hoff W (2002) Water transport in brick, stone and concrete. Spon Press, London and New York

International Energy Agency (1990) Guidelines and Practice. In: Energy conservation in buildings and community systems programme, Annex XIV consensation and evergy, vol 2

International Union of Pure and Applied Chemistry (1994) Recommendations for the characterization of porous solids. Pure Appl Chem 66(8):1739–1758

Karoglou M, Moropoulou A, Giakoumaki A, Krokida MK (2005a) Capillary rise kinetics of some building materials. J Colloid Interface Sci 284(1):260–264

Karoglou M, Moropoulou A, Krokida MK, Maroulis ZB (2005b) Drying kinetics of some building materials. Dry Technol 23(1–2):305–315

Karoglou M, Moropoulou A, Krokida MK, Maroulis ZB (2005c) Water sorption isotherms of some building materials. Dry Technol 23(1–2):289–303x

Kieß K (1989) Bauphysikalische Einflüsse bei der Krustenbildung am Gestein alter Bauwerke. Bauphysic 11(1):44–49

Konstantopoulos DK, Maroulis VZ, Karoglou M, Moropoulou A, Mujumdar AS (2007) HygroScope: a moisture transfer simulator for buildings. Dry Technol 25(6):1119–1125

Mangel A (1999) Investigating a range of solid samples by automatic water sorption. J Therm Anal Calorim 55:581–599

Künzel HM (1995) Simultaneous heat and moisture transport in building components:one- and two-dimensional calculation using simple parameters. In: Phd thesis, Fraunhofer IRB Verlag

Marinos-Kouris D, Maroulis ZB (1995) Transport properties in the drying of solids. In: Mujumdar AS (ed) Handbook of industrial drying, 2nd edn. Marcel Dekker, New York

Massari G, Massari I (1981) Risanamento igienico dei locali umidi. Hoepli, Trento

Meng B (1994) Calculation of moisture transport coefficients on the basis of relevant pore structure parameters. Mater Struct 27:125–134

Oliver A (1997) Dampness in buildings. In: Douglas J, Sterling JS (eds) 2nd edn. Blackwell science, Great Britain

Oxley TA, Gobert EG (1998) Dampness in buildings: diagnosis, treatment, instruments, 2nd edn. Biddles Ltd Guilford and King's Lynn, Great Britain

Saravacos GD, Maroulis ZB (2001) Transport properties of foods. Marcel Dekker, New York

Torraca G (1981) Porous building materials. ICCROM, Rome

Washburn E (1921) The dynamics of capillary flow. Phys Rev 17(3):273

Zaknoune A, Glouannec P, Salagnac P (2012) Estimation of moisture transport coefficients in porous materials using experimental drying kinetics. Int J Heat Mass Transfer 48:205–215

[illegible]

# Chapter 3
# Wetting and Drying Kinetics of Building Materials

**Eva Barreira, João M. P. Q. Delgado and Vasco Peixoto de Freitas**

**Abstract** Moisture is one of the most deteriorating factors of buildings. The moisture content depends on hygroscopic equilibrium between buildings materials and environment, which is determined by the drying and wetting rate of masonry. So, the moisture content is not only determined by the water that is absorbed by the material, but also by the amount of water that is evaporated under favourable conditions, which is described by the drying process. This work presents the results of an experimental evaluation of the drying kinetics, considering both surface and bulk moisture transfer. Two different specimens were used: External Thermal Insulation Composite Systems (ETICS) to study surface moisture transfer and solid brick to analyse the bulk moisture transfer. For both samples the drying kinetics was assessed considering different environment conditions (air temperature and humidity). Six different first-order kinetics models, available in the literature, were adjusted to describe the drying process and estimate the equilibrium moisture content of ETICS and solid red brick samples. The results point that Hii et al. and Midilli models allow the best fit and that the drying time constant is strongly affected by the moisture transfer phenomenon (at surface or bulk) and by the drying air conditions. It was also estimated the apparent molecular diffusion coefficient for solid red brick samples and its variation with temperature.

**Keywords** Wetting • Drying • Hygrothermal behaviour • Drying kinetics • Diffusion coefficients • ETICS systems • Solid red brick

E. Barreira · J. M. P. Q. Delgado (✉) · V. P. de Freitas
LFC—Building Physics Laboratory, Faculty of Engineering, Civil Engineering Department, University of Porto, 4200-465 Porto, Portugal
e-mail: jdelgado@fe.up.pt

E. Barreira
e-mail: barreira@fe.up.pt

V. P. de Freitas
e-mail: vpfreita@fe.up.pt

J. M. P. Q. Delgado (ed.), *Drying and Wetting of Building Materials and Components*, Building Pathology and Rehabilitation 4, DOI: 10.1007/978-3-319-04531-3_3,

## Nomenclature

| | |
|---|---|
| $a.\ b.\ n$ | Empirical coefficients in model |
| $A$ | Area |
| $h$ | Initial sorption rate |
| $K$ | Drying kinetic coefficient |
| $MR$ | Moisture ratio. $MR = (w - w_{eq})/(w_0 - w_{eq})$ |
| $n$ | Order of kinetic equations |
| $N$ | Number of experimental points |
| $p$ | Number of estimated parameters |
| $P$ | Water vapour pressure |
| $S$ | Standard deviation |
| $t$ | Time |
| $t_c$ | Drying time constant |
| $t_{0.5}$ | Time to reach 50 % of equilibrium moisture content |
| $T$ | Temperature |
| $w$ | Moisture content |
| $w_{eq}$ | Equilibrium moisture content |
| $w_0$ | Initial moisture content |
| $w*$ | Predicted value of moisture content |
| $\beta$ | Superficial diffusion coefficient |

## 3.1 Introduction

External thermal insulation composite systems (ETICS) for walls—expanded polystyrene insulation faced with a thin rendering—have been used in various countries since the mid twentieth century. In Portugal, these systems were first applied in the nineties. Unfortunately, the results weren't always as expected. In recent years, stains have frequently appeared on facades covered with ETICS as a result of algae and mildew growth. These stains compromise the facades' aspect and have been greatly contested by property owners. The thermal advantages of ETICS are, therefore, undervalued by owners due to the conspicuous degradation. This problem is intensified by poor building maintenance habits in Portugal. It is essential to study the causes of degradation in order to understand the hygrothermal behaviour of facades covered with ETICS (see Barreira and Freitas 2005).

The drying of built-in moisture can cause several damages in building components, especially if materials with low vapour permeability are applied in the adjacent layers. For that reason, it is very important to estimate the drying time of the material before its application. Solid red bricks are commonly used in the building industry in Europe and therefore it is important to understand its behaviour regarding the drying process.

Moisture is one of the most deteriorating factors of buildings, and the investigation of materials drying kinetics is very useful in understanding the effect of environmental conditions on the moisture content of building materials. In literature it is possible to find many moisture transfer models used to describe the drying process of porous materials (Maroulis and Saravacos, 2003). For example, Derdour et al. (2000) have studied the effect of air temperature, relative humidity, and air velocity on the drying time constant of building materials. Krus and Holm (1999) developed a simple method to approximate the liquid transport coefficients describing the adsorption and drying of some building materials. Finally, Bednar (2002) investigated the influence of building materials size, material insulation, and climate conditions on the liquid moisture transport coefficient during drying experiments.

The objective of this work is to investigate the drying kinetics of two different specimens: ETICS to study surface moisture transfer and solid brick samples to analyse the bulk moisture transfer. A first-order and a second-order drying kinetic models are used to describe mass transfer phenomenon. Five different first-order kinetics models, available in the literature (linear type, power type, exponential type, Arrhenius type and Logarithmic type) were tested to assess the better fit to describe the drying process and estimate the equilibrium moisture content. The parameters of the models that better describe the phenomenon were found to be strongly affected by the moisture transfer (surface or bulk) and the drying air conditions.

Many moisture transfer models have been used to describe the drying process of porous materials in the food and chemical industry (Maroulis and Saravacos 2003). The objective of this work is to investigate the drying kinetics of solid red brick samples by the use of different kinetic models, such as linear type, power type, exponential type, Arrhenius type and Logarithmic type. The parameters of the proposed model were found to be affected strongly by the material and the drying air conditions.

It is, therefore, the objectives of this study were:

- To observe the effect of drying temperature during drying process of solid red brick samples;
- To estimate the constant of selected model equations, diffusion coefficient and activation energy;
- To determine the effect of temperature on constants and coefficients in the selected models and diffusion coefficient.

## 3.2 Theory

In the vast majority of building materials, namely specimens with a stable pore structure, the drying process, mass variation in function of time, is well represented by the solution of first or second order kinetics equations:

$$\frac{dw}{dt} = \frac{1}{t_{c1}}(w_{\text{eq}} - w) \tag{3.1}$$

where $dw/dt$ is the drying rate, $w$ is the material moisture content at time $t$, $t_{c1}$ is the drying time constant and $w_{eq}$ is the equilibrium material moisture content (see Delgado et al. 2006). The drying time constant $t_c$ is used instead of the well-known drying constant, $K$, $t_c = 1/K$. The drying time constant is consistent with the process control terminology for first-order dynamic systems and its value has physical meaning denoting the required time to remove two thirds of the total moisture.

When drying conditions are kept constant, the integration of Eq. (3.1) for the initial and end conditions $t = 0$ to $t = t$ and $w = w_0$ to $w = w$, gives

$$w = w_{eq} + (w_0 - w_{eq}).\exp\left(-\frac{t}{t_{c1}}\right) \tag{3.2}$$

In order to fit Eq. (3.2) to experimental data, the equilibrium desorption capacity, $w_{eq}$, should be known. In most cases from the literature, the first order equation does not fit well over the range of contact times under investigation. Furthermore, one has to find some means of extrapolating the experimental data to $t \to \infty$, or treat $w_{eq}$ as an adjustable parameter to be determined by trial and error. For this reason, it is necessary to use a trial and error solution method to obtain the equilibrium sorption capacity, $w_{eq}$.

Another model for the analysis of sorption kinetics is the second-order model and the rate law for this system is expressed as

$$\frac{dw}{dt} = \frac{1}{t_{c2}}(w_{eq} - w)^2 \tag{3.3}$$

Integrating Eq. (3.3), for the boundary conditions $t = 0$ to $t = t$ and $w = w_0$ to $w = w$ gives

$$\frac{1}{w_{eq} - w} = \frac{1}{w_{eq} - w_0} - \frac{t}{t_{c2}} \tag{3.4}$$

where $t_{c2}$ is the pseudo-second-order drying time constant. Equation (3.4), can be rearranged to obtain a linear form,

$$\frac{t}{w - w_0} = a.t - b \tag{3.5}$$

with $a = 1/(w_{eq} - w_0)$ (slope) and $b = t_{c2}/(w_{eq} - w_0)^2$ (intercept). The plot of $t/(w - w_0)$ *vs.* $t$ gives a straight line (see Fig. 3.1) and the equilibrium water content and sorption rate constant could be evaluated from the slope ($w_{eq} = \tan^{-1}\theta$) and intercept, respectively.

However, there are some building materials whose mass variation in function of time is not well represented by the solution of first or second order sorption kinetics equations. In these situations, it is necessary to use the general form of Eq. (3.1),

$$\frac{dw}{dt} = \frac{1}{t_{cn}}(w_{eq} - w)^n \tag{3.6}$$

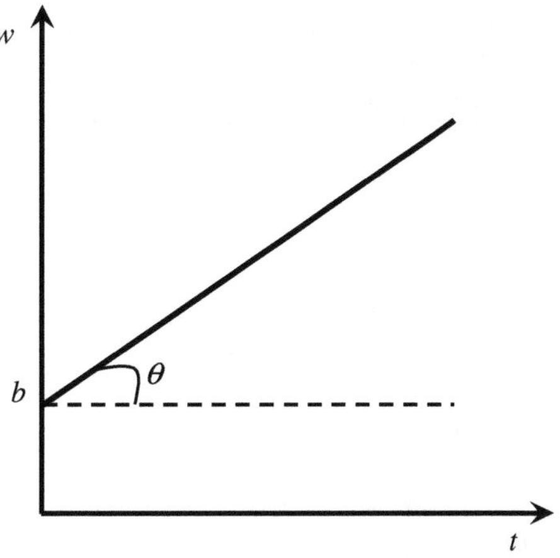


**Fig. 3.1** Dependence of $t/(w - w_0) = f(t)$ for the drying kinetics model

and the integration of Eq. (3.6) with the same boundary conditions of Eq. (3.1) gives:

$$\frac{1}{(n-1).\left(w_{\text{eq}} - w\right)^{n-1}} = \frac{1}{(n-1).\left(w_{\text{eq}} - w_0\right)^{n-1}} - \frac{t}{t_{cn}} \tag{3.7}$$

the general solution of Eq. (3.8), valid for $n > 1$.

Interesting parameters obtained by sorption kinetic equations are the initial drying rate, $h$, and the time to reach 50 % of initial water content, $t_{0.5}$, given by

$$h = \left(\frac{dw}{dt}\right)_{t=0} = \frac{\left(w_{\text{eq}} - w_0\right)^n}{t_{cn}} \tag{3.8}$$

$$t_{0.5} = \frac{2^{n-1}. - 1}{(n-1).\left(w_{\text{eq}} - w_0\right)^{n-1}}.t_{cn} \text{ (valid for } n > 1) \tag{3.9}$$

Equations (3.2) and (3.4) calculate the required drying time $t$ to remove moisture from the initial, $w_0$, to the final, $w$, material moisture content. The model parameters are estimated by minimizing the following standard deviation,

$$S = \left[\sum_{i=1}^{N} \frac{(w - w*)^2}{N - p}\right]^{1/2} \tag{3.10}$$

where $w$ and $w^*$ are the experimental and predicted values of material moisture content, $N$ is the number of experimental points and $p$ is the number of estimated parameters.

Finally, the drying curves were fitted with different empirical and semi-empirical drying models (see Table 3.1). Regression analyses of these equations were done by

**Table 3.1** Mathematical models (semi-theoretical and empirical) given by various authors for drying curves

| Model equation | Model name |
|---|---|
| $MR = e^{-t/t_c}$ | Newton equation |
| $MR = e^{-t^n/t_c}$ | Page equation |
| $MR = e^{-(t/t_c)^n}$ | Modified page equation |
| $MR = a \times e^{-t/t_c}$ | Henderson and Pabis |
| $MR = a \times e^{-t/t_{c1}} + b \times e^{-t/t_{c2}}$ | Two-term |
| $MR = a \times e^{-t/t_{c1}} + (1 - a) \times e^{-t/t_{c2}}$ | Diffusion approach |
| $MR = a \times e^{-t/t_{c1}} + b \times e^{-t/t_{c2}} + c \times e^{-t/t_{c3}}$ | Henderson and Pabis modified |
| $MR = a \times e^{-t/t_c} + c$ | Logarithmic equation |
| $MR = a \times e^{-t^n/t_c} + b \times t$ | Midilli equation |
| $t = a \ln (MR) + b[\ln (MR)]^2$ | Thompson |
| $MR = 1 + at + bt^2$ | Wang and Singh |
| $MR = e^{-(t.t_c)^n}$ | Welbull distribution |
| $MR = e^{(-t/t_{c1}/(1+t/t_{c2}))}$ | Aghbashlo et al |
| $MR = (a + t/t_c)^2$ | Vega and Lemus |
| $MR = a.e^{-(t^n/t_{c1})} + b.e^{-(t^n/t_{c2})}$ | Hii et al |

using STATISTICA routine. The regression coefficient ($R^2$) was primary criterion for selecting the best equation to describe the drying curves of ETICS. The performance of derived new models was evaluated using various statistical parameters such as the mean bias error (*MBE*), the root mean square error (*RMSE*) and chi-square ($\chi^2$), in addition to the regression coefficient ($R^2$). These parameters can be calculated as following:

$$MBE = \frac{1}{N}\sum_{i=1}^{N}\left(MR_{pre,i} - MR_{\exp,i}\right) \quad (3.11a)$$

$$RMSE = \sqrt{\frac{1}{N}\sum_{i=1}^{N}\left(MR_{pre,i} - MR_{\exp,i}\right)} \quad (3.11b)$$

$$\chi^2 = \frac{\sum_{i=1}^{N}\left(MR_{\exp,i} - MR_{pre,i}\right)^2}{N - n} \quad (3.11c)$$

where *MR* is the moisture ratio, $MR = \frac{w(t) - w_{eq}}{w_0 - w_{eq}}$.

The variation of the constants and coefficients in these equations with temperature and relative humidity can be described by several different models. These are generally linear type, power type, exponential type, Arrhenius type and Logarithmic type. By virtue of this consideration, numerous empirical equations can be obtained from empirical and semi-empirical models. In this study, the constant and coefficient of models in Table 3.1 were written as power and exponential function of temperature and relative humidity.

The experimental drying data for determination of diffusivity was interpreted by using Fick's diffusion model. In a one-dimensional formulation with the diffusing substance moving in the direction normal to a sheet of medium of thickness $L$, the diffusion equation can be written as (Crank 1975)

$$\frac{\partial w}{\partial t} = D_{eff} \frac{\partial^2 w}{\partial x^2} \tag{3.12}$$

subject to the following boundary conditions

$$t = 0 \quad 0 < x < \infty \quad C = C_{\rm i} \tag{3.13a}$$

$$t > 0 \quad x = 0 \quad C = C_{\rm eq} \tag{3.13b}$$

$$t > 0 \quad x \to \infty \quad C = C_{\rm i} \tag{3.13c}$$

The analytical solution of Eq. (3.12) with the initial and boundary conditions (3.13a–13.3c) is

$$\frac{C - C_{\rm i}}{C_{\rm eq} - C_{\rm i}} = 1 - erf\left(\frac{x}{2\sqrt{D_{\rm m}t}}\right) \tag{3.14}$$

Integrating in respect of $t$ the rate of the penetration of sample face unit area ($x = 0$) by water vapour, the total amount of diffusing substance in time $t$ is obtained, then the appropriate solution of the diffusion Eq. (3.12) may be written as follows,

$$MR = \frac{8}{\pi^2} \times \sum_{n=0}^{\infty} \frac{1}{(2n+1)^2} \exp\left[-(2n+1)^2 \pi^2 \frac{D_{eff} t}{L^2}\right] \tag{3.15}$$

and for long drying times (neglecting the higher order term by setting $n = 0$) it has been simplified as following:

$$\ln(MR) = \ln\left(\frac{8}{\pi^2}\right) - \left[\pi^2 \frac{D_{eff} t}{L^2}\right] \tag{3.16}$$

Finally, the effect of temperature on diffusivity is described using Arrhenius relationship:

$$D_{eff} = A_0 \times \exp\left(-\frac{E_a}{RT}\right). \tag{3.17}$$

**Table 3.2** Materials used in the ETICS system

| ETICS components | $b$ (mm) | $\rho$ (kg/m$^3$) | $\varepsilon$ (m$^3$/m$^3$) |
|---|---|---|---|
| Expanded polystyrene | 60 | 20 | 0.95 |
| Base coat | 2 | 1.55 | – |
| Finishing coat—rough | 1 | 180 | 0.12 |
| Finishing coat—smooth | 1 | 160 | 0.12 |

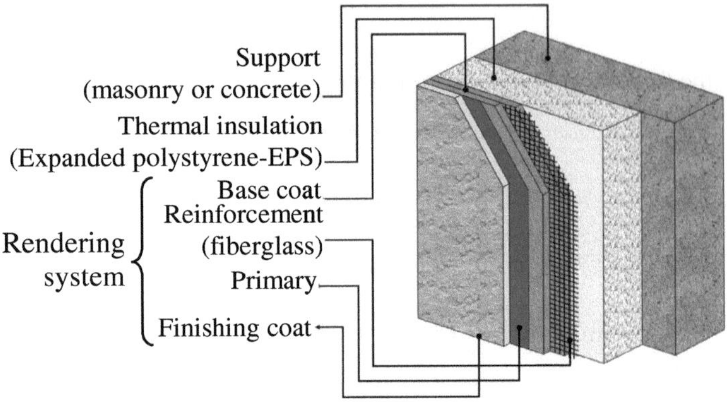


**Fig. 3.2** Configuration of ETICS samples

## 3.3 Experimental Setup

### *3.3.1 Drying Kinetics of ETICS Systems*

In these tests, two types of External Thermal Insulation Composite Systems (ETICS) specimens, smooth and rough, were prepared according to manufacturer's specifications. The material characteristics are summarized in Table 3.2 and sketched in Fig. 3.2.

Experiments to determine the influence of process variables on the drying kinetics were performed at constant conditions of relative humidity and temperature. The drying experiments were carried out on sample size of $0.30 \times 0.30$ m$^2$, each material. Before the start of the experiment, the samples were saturated with water through immersion in distilled water.

To perform these experiments, a climatic chamber was used (see Fig. 3.3) to subject the specimens to simulated climatic conditions over extended periods of time; namely the control of temperature and relative humidity. Both the relative humidity and the temperature of the chamber can be independently controlled to constant values ($T$ = 25 °C and $RH$ = 50 %). A precision balance was located inside the climatic chamber and the mass change registered continuously by a personal computer.

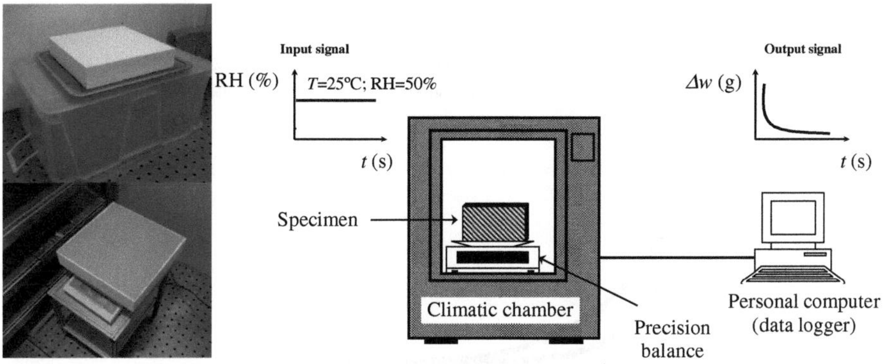


**Fig. 3.3** Sketch of the experimental set-up

### *3.3.2 Drying Kinetics of Solid Red Brick Samples*

The drying experiments were carried out on sample size of approximately $5.5 \times 5.5 \times 11$ cm$^3$. Experiments to determine the influence of process variables on the drying kinetics were performed at constant conditions of relative humidity and temperature. Before the start of the experiment, the samples were saturated with water.

Each sample was analyzed at four levels of air-temperature (15, 20, 25 and 30 °C) and three levels of relative humidity of the air (50, 65 and 80 %).

To perform these experiments, a climatic chamber was used to subject the specimens to simulated climatic conditions over extended periods of time; namely to control the temperature and relative humidity. Both parameters can be independently settled to constant values. A precision balance was located inside the climatic chamber and the mass change registered manually.

## 3.4 Results and Discussion

### *3.4.1 Drying Kinetics of ETICS Systems*

The series of tests performed on the two types of ETICS specimens meant to analyze the response in terms of water vapour content variation during the drying process at constant conditions of relative humidity and temperature. Figures 3.4a shows the mass variation observed in the two samples analysed.

An additional way of characterizing the experimental results was the application of kinetics models equations to desorption phase. Based on the performed analysis, it was found that the drying process of water vapour onto these samples proceeds in accordance with the pseudo second-order model. This is confirmed by rectilinear

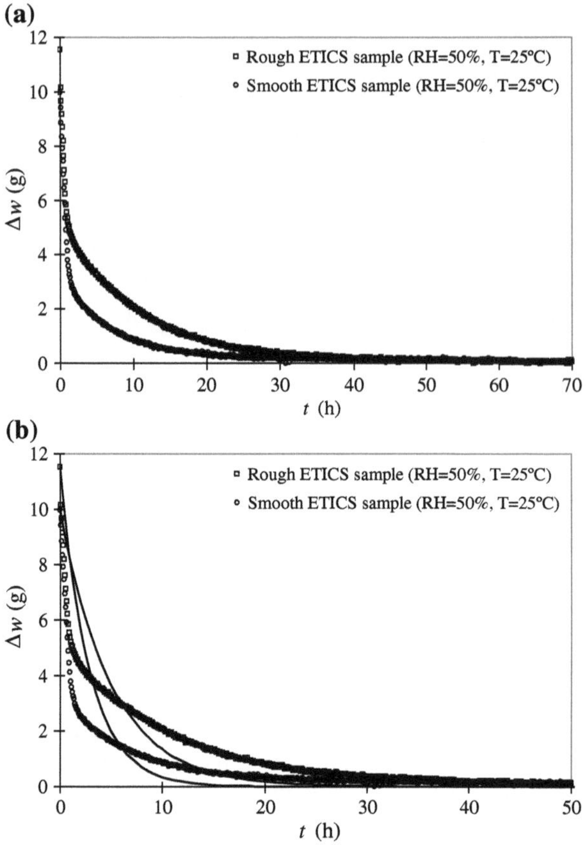


**Fig. 3.4** **a** Example of mass variation observed during the drying process. **b** Drying kinetics using the first-order model, Eq. (3.2)

diagrams of sorption showed in Figs. 3.5a. The experimental values do not deviate from the lines by more than 10 % and the correlation coefficients were ~0.97.

The kinetics parameters of the water vapour desorption on building materials tested are listed in Table 3.3, including the time to reach 50 % of equilibrium water content, the equilibrium water content and theoretical kinetic parameters as kinetic coefficient (*K*) and initial drying rate (*h*). The experimental curves, obtained from experimental data, of desorption are show in Figs. 3.4b and 3.5b. An advantage of the use of kinetics models for drying processes is the possibility that the end of a drying process, $w_{eq}$, may be predicted, using only a few hours experiment, with good agreement of the results. Another, important information obtained by the use of kinetics sorption models is given by the initial drying rate value, *h*. The *h* value gives a measure that can be a precious help in the characterization of initial hygroscopic performance of different building materials.

Finally, drying curves were fitted with five different empirical and semi-empirical drying models (see Table 3.4). The model number, model constant or coefficient and standard deviation error of five different models used for moisture ratio, *MR*, change

Fig. 3.5 **a** Dependence of $t/(w - w_0) = f(t)$. **b** Drying kinetics with 2nd-order model, Eq. (3.4)

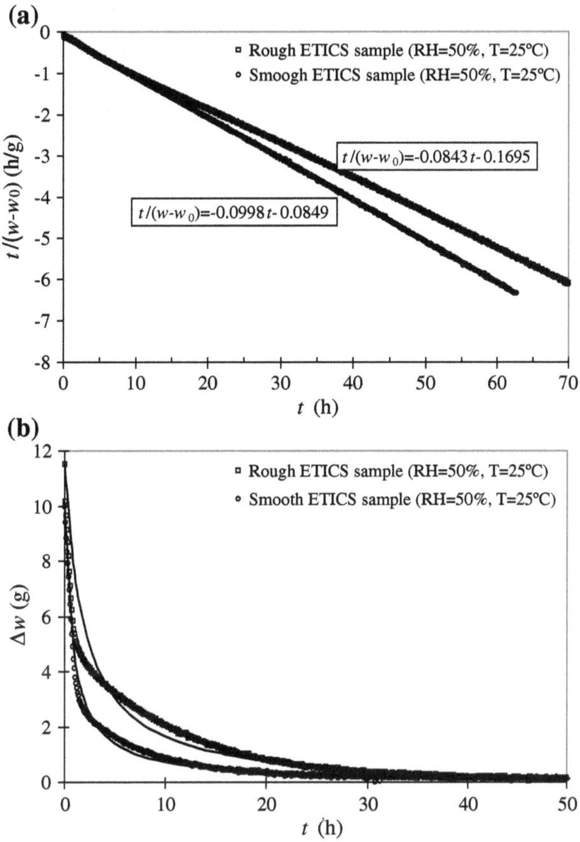


**Table 3.3** Drying kinetic parameters obtained by two different models

| Materials | | $T$ (°C) | RH (%) | $w_0$ (g) | $t_{c1}$ (h) | $K$ ($h^{-1}$) | $t_{c2}$ (h.g) | $t_{0.5}$ (h) | $h$ (g/h) | $g$ (g/h) |
|---|---|---|---|---|---|---|---|---|---|---|
| ETICS | Rough | 25 | 50 | 11.53 | 2.86 | 0.35 | 23.81 | 2.06 | 5.58 | – |
| | Smooth | 25 | 50 | 9.97 | 5.00 | 0.20 | 8.55 | 0.86 | 11.62 | 11.34 |

with time are presented in Table 3.4. The criterion used for model selection was magnitude of standard deviation error for each model. Models 2 (Page equation) and 5 (Midilli et al. equation) have the best standard deviation error. Therefore, Models 2 and 5 can be proposed to evaluate the moisture ratio of ETICS drying process (see Fig. 3.6).

## 3.4.2 Drying Kinetics of Solid Red Brick Samples

Figure 3.7 shows the variation of the moisture ratio (*MR*) with time during the drying process, at different temperatures and relative humidity values, for solid red brick samples.

**Table 3.4** Mathematical models given by various authors for drying curves and result of statistical analyses on the modelling of moisture content and drying time

| Model No. | Name | Model equation and references | Model constants | | Error (%) | |
|---|---|---|---|---|---|---|
| | | | Rough ETICS | Smooth ETICS | Rough | Smooth |
| 1 | Newton | $MR = \exp(-k.t)$ (Liu and Bakker-Arkema 1997) | $k = 0.259$ | $k = 0.545$ | 0.48 | 0.21 |
| 2 | Page | $MR = exp(-k.t^n)$ (White and Ross 1981) | $k = 0.612$<br>$n = 0.485$ | $k = 0.808$<br>$n = 0.495$ | 0.03 | 0.04 |
| 3 | Page modified | $MR = exp(-(k.t)^n)$ (White and Ross 1981) | $k = 0.259$<br>$n = 1.000$ | $k = 0.545$<br>$n = 1.000$ | 0.48 | 0.21 |
| 4 | Logarithmic | $MR = a.\exp(-k.t) + b$ (Yagcıoglu et al. 1999) | $k = 0.133$<br>$a = 0.570$<br>$b = 0.016$ | $k = 0.429$<br>$a = 0.757$<br>$b = 0.027$ | 0.09 | 0.10 |
| 5 | Midilli et al. | $MR = a.\exp(-k.t^n) + b.t$ (Midilli et al. 2002) | $k = 0.652$<br>$n = 0.447$<br>$a = 1.017$<br>$b = -3.4 \times 10^{-4}$ | $k = 0.932$<br>$n = 0.472$<br>$a = 1.154$<br>$b = 2.1 \times 10^{-4}$ | 0.02 | 0.03 |

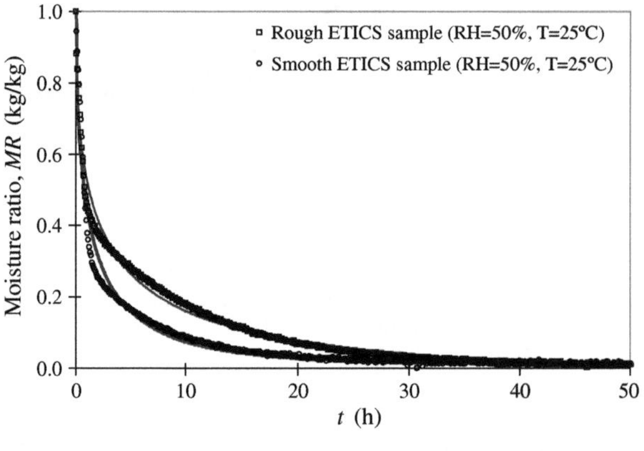


**Fig. 3.6** Experimental and predicted *MR* (Midilli et al. equation) versus drying time

Drying rate decreases continuously with time and decreasing moisture content. This result shows that dominant physical mechanism governing moisture movement in drying process of solid red brick samples is the second drying stage (for ETICS is the first drying stage). The diffusion within the building material limits the rate of evaporation during the second stage of the drying process.

Figures 3.7, 3.8 and 3.9 show the changes in the moisture ratio (*MR*) with time during the drying process at different temperatures and relative humidity's. Drying rate decreases continuously with time and decreasing moisture content.

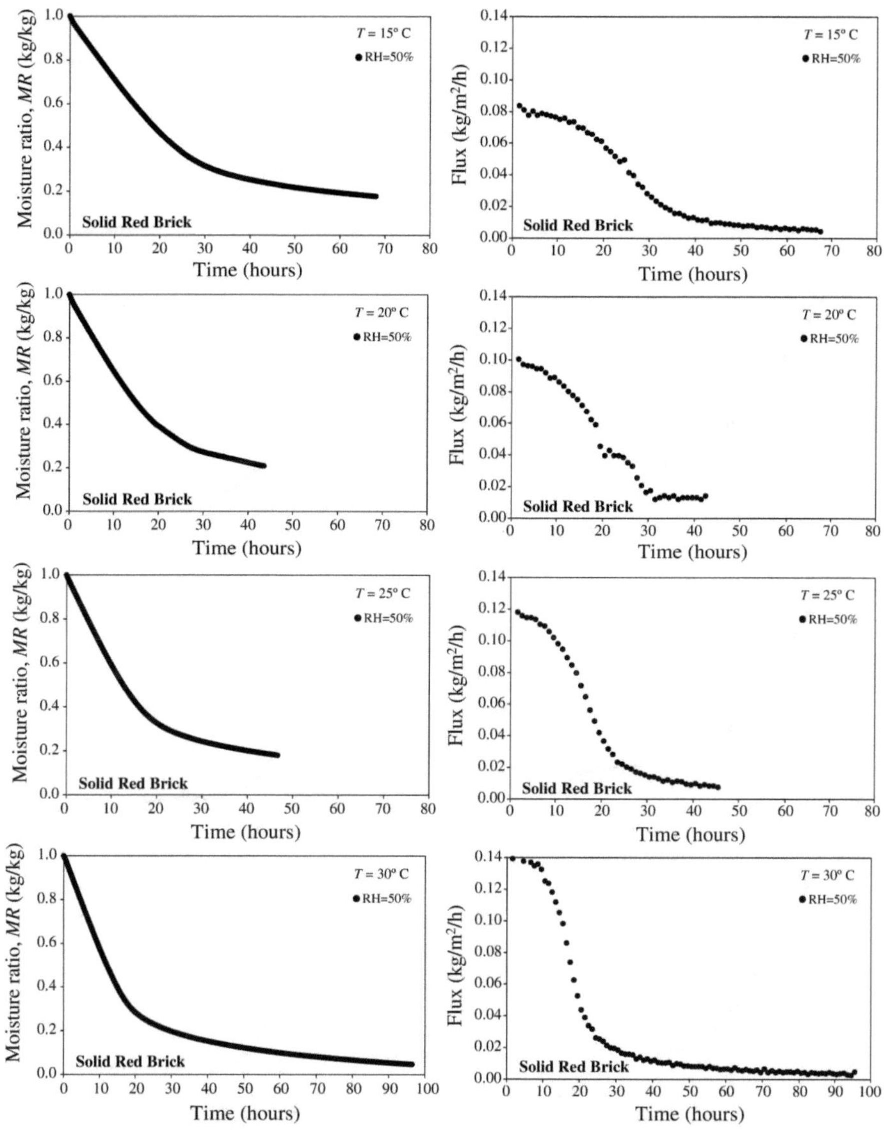


**Fig. 3.7** Experimental MR and Flux values *versus* drying time, at RH = 50 % and different temperatures

As indicated in this curves, the whole experiment in this study came true in falling rate period.

The average drying rate values between 15 and 30 °C drying temperatures were determined and these values increased when the drying temperature was increased from 15 to 30 °C, for both samples.

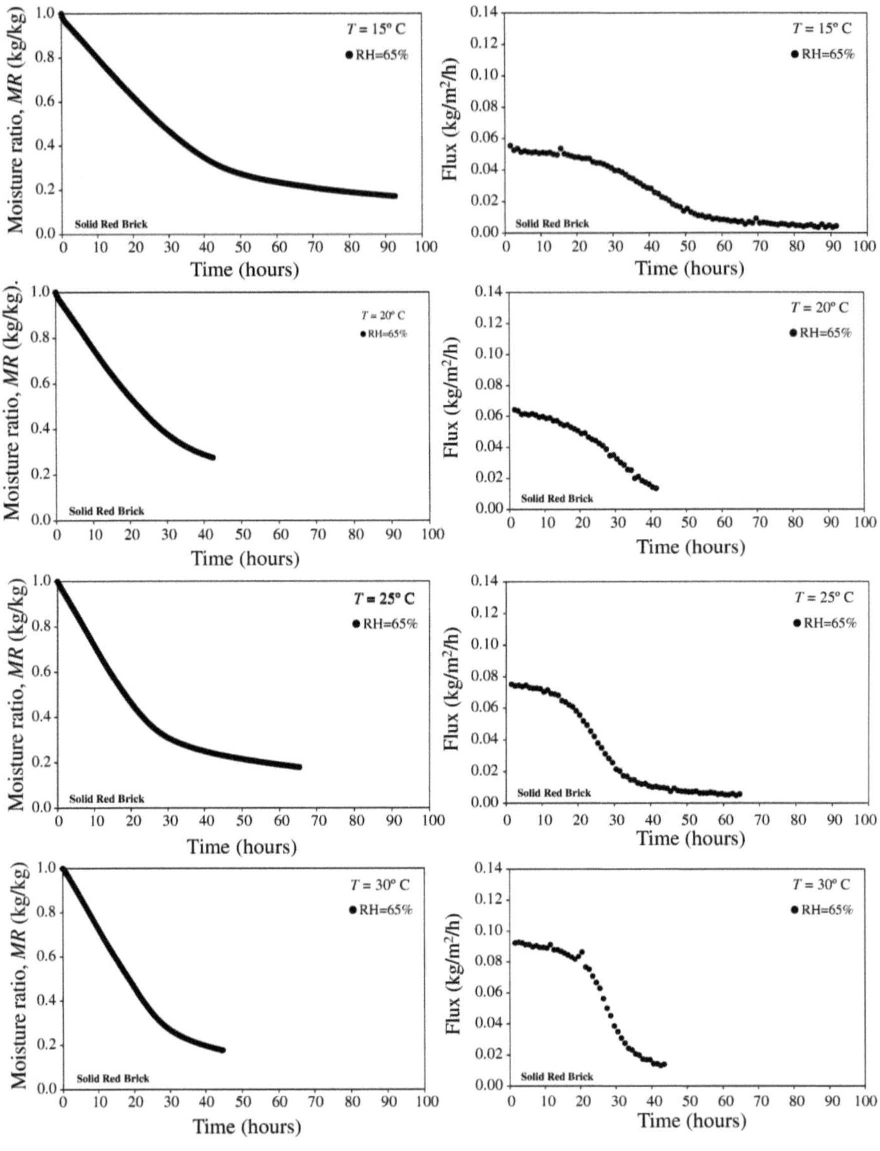


**Fig. 3.8** Experimental MR and Flux values *versus* drying time, at RH = 65 % and different temperatures

The common feature of the curves presented in Fig. 3.10 is very similar to a typical drying curve, with the moisture ratio decreasing exponentially with time. The same moisture content (approximately MR = 0) was obtained in different drying time for each drying temperatures and relative humidity's.

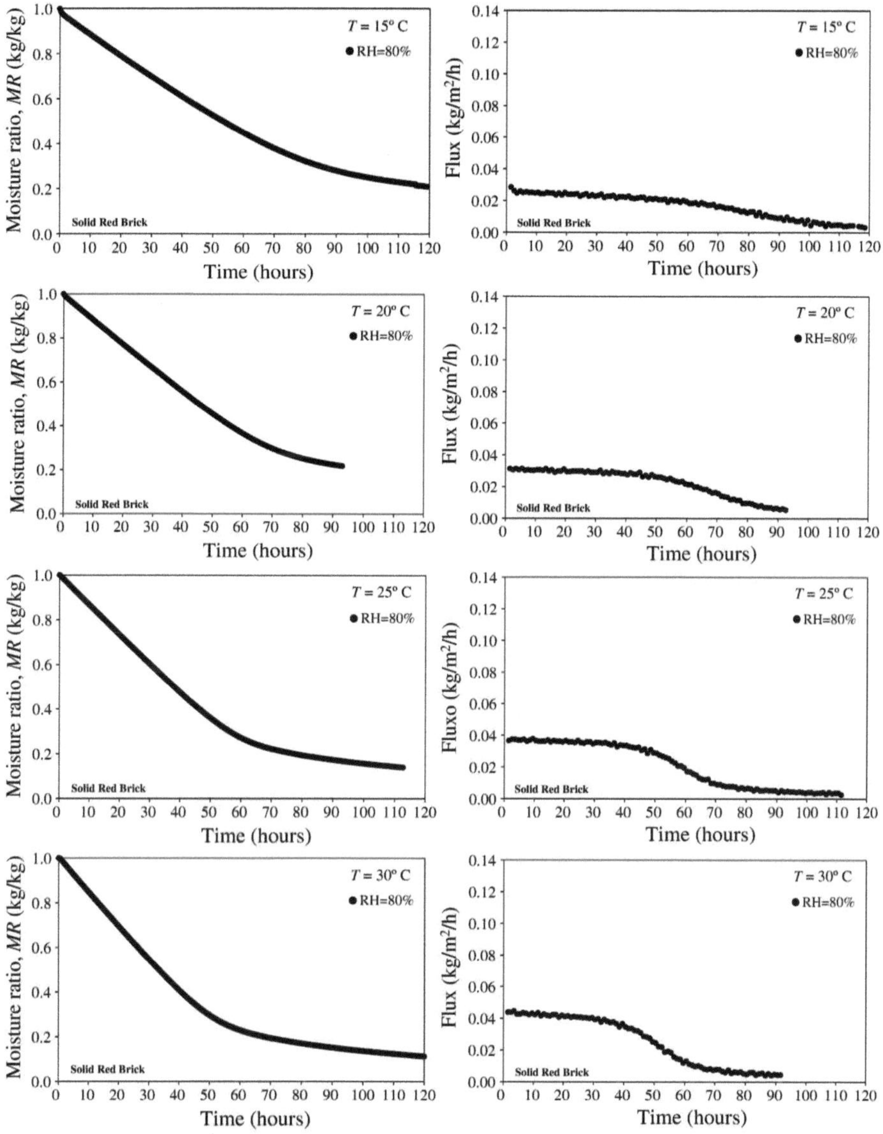


**Fig. 3.9** Experimental MR and Flux values *versus* drying time, at RH = 80 % and different temperatures

Six drying models have been used to describe drying curves. These models are presented in Tables 3.5 and 3.6. The criterion used for selecting the models that better describe the drying process was the magnitude of the relative error for each one. The Hii et al. and Midilli models presented the best results for the solid red brick samples, such as for ETICS.

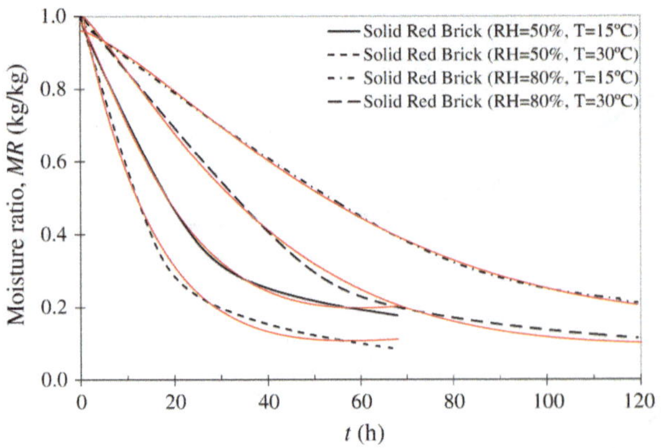


**Fig. 3.10** Experimental and predicted MR versus drying time, for solid red brick samples

**Table 3.5** Results of the statistical analyses (chi-square, $\chi 2$) with the drying models tested time, for solid red brick samples

| Sample | RH (%) | T(°C) | Error (%) | | | | | |
|---|---|---|---|---|---|---|---|---|
| | | | Newton | Page | Logarithmic | Midilli | Henderson–Pabis | Hii et al. |
| RB–I | 15 | 50 | 125.21 | 81.53 | 19.49 | 6.56 | 116.14 | 1.97 |
| RB–II | 20 | 50 | 32.24 | 24.24 | 8.54 | 1.00 | 30.80 | 0.86 |
| RB–III | 25 | 50 | 94.34 | 59.27 | 11.88 | 4.8 | 87.6 | 1.14 |
| RB–IV | 30 | 50 | 228.1 | 140.01 | 35.07 | 48.5 | 219.58 | 2.06 |
| RB-V | 15 | 65 | 93.06 | 79.49 | 34.15 | 4.80 | 91.49 | 3.16 |
| RB–VI | 20 | 65 | 9.44 | 4.35 | 6.25 | 1.02 | 6.46 | 0.96 |
| RB–VII | 25 | 65 | 115.32 | 86.88 | 23.14 | 6.79 | 112.37 | 2.13 |
| RB–VIII | 30 | 65 | 55.35 | 14.61 | 22.76 | 3.02 | 26.42 | 2.16 |
| RB–IX | 15 | 80 | 42.76 | 23.1 | 105.67 | 4.69 | 30.24 | 2.38 |
| RB–X | 20 | 80 | 90.25 | 10.19 | 21.69 | 5.14 | 49.13 | 2.36 |
| RB–XI | 25 | 80 | 131.1 | 75.67 | 81.2 | 8.13 | 81.56 | 8.02 |
| RB–XII | 30 | 80 | 228.43 | 227.65 | 85.71 | 38.25 | 216.14 | 8.6 |
| Average | | | 103.81 | 68.92 | 37.96 | 11.06 | 88.99 | 2.96 |

Table 3.6 shows the drying time constant and drying constant obtained at different temperatures and relative humidity values.

In an attempt to simplify the mode of water sorption by food materials Peleg (1988) proposed, a two parameter, nonexponential empirical equation, which became known as the Peleg's model

$$MR = 1 - \frac{t}{k_1 + t.k_2} \tag{3.18}$$

where $k_1$ is the Peleg rate constant, and $k_2$ is the Peleg capacity constant. The units of $k_1$ and $k_2$ will correspond to those of the moisture units.

**Table 3.6** Values of drying time constant and drying constant ($K = 1/t_c$) for the best fitted models tested (solid red brick samples)

| Sample | RH (%) | T (°C) | $K$ ($h^{-1}$) | | $t_c$ (h) | |
|---|---|---|---|---|---|---|
| | | | Hii et al. | Midilli | Hii et al. | Midilli |
| RB–I | 15 | 50 | 0.017 | 0.024 | 59.26 | 41.40 |
| RB–II | 20 | 50 | 0.028 | 0.030 | 35.25 | 33.29 |
| RB–III | 25 | 50 | 0.030 | 0.044 | 33.26 | 22.87 |
| RB–IV | 30 | 50 | 0.002 | 0.060 | 487.08 | 16.73 |
| RB–V | 15 | 65 | 0.009 | 0.012 | 116.83 | 86.96 |
| RB–VI | 20 | 65 | 0.017 | 0.017 | 60.14 | 57.36 |
| RB–VII | 25 | 65 | 0.015 | 0.024 | 66.42 | 41.20 |
| RB–VIII | 30 | 65 | 0.013 | 0.012 | 77.97 | 85.09 |
| RB–IX | 15 | 80 | – | 0.004 | – | 245.09 |
| RB–X | 20 | 80 | 0.005 | 0.006 | 220.76 | 168.04 |
| RB–XI | 25 | 80 | 0.005 | 0.004 | 197.91 | 245.45 |
| RB–XII | 30 | 80 | 0.004 | 0.011 | 250.80 | 91.20 |

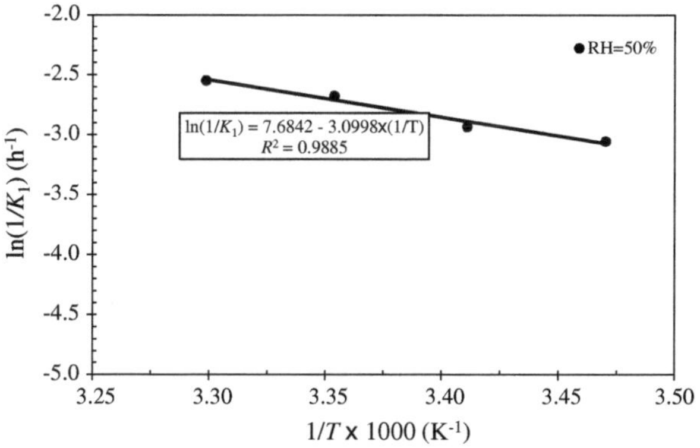


**Fig. 3.11** Effect of drying temperature on drying rate constant, for solid *red* brick samples at 25 °C and RH = 50 %

Finally, the relationships between $1/k_1$ and the drying temperature could be expressed by linearized Arrhenius equation as followed:

$$\ln\left(\frac{1}{k_1}\right) = \ln k_0 - \frac{E_a}{R.T} \tag{3.19}$$

where $k_0$ is a constant (h), $E_a$ is activation energy (kJ/mol), $R$ is universal gas constant (8.314 kJ/mol K), and $T$ is the absolute temperature (K).

Figures 3.11, 3.12 and 3.13 shows Arrhenius plot of initial drying rate and the average activation energy for solid red brick samles at 25 °C and different relative humidity's. The $E_a$ values suggest that the prevailing water transport is diffusion ($E_a$ values between 21.13 and 33.80 kJ/mol, at 25 °C and different relative humidity's).

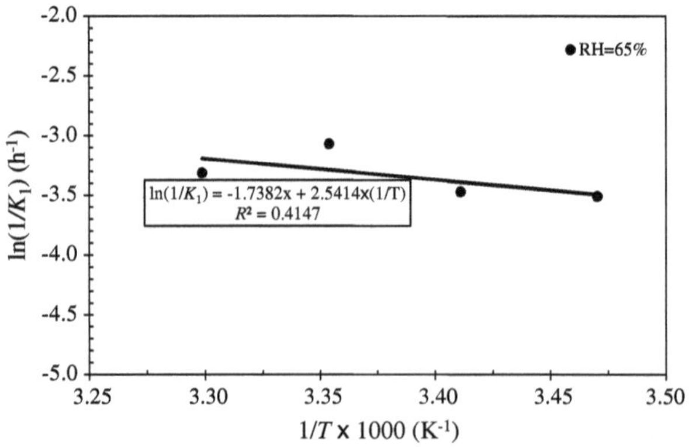


**Fig. 3.12** Effect of drying temperature on drying rate constant, for solid *red* brick samples at 25 °C and RH = 60 %

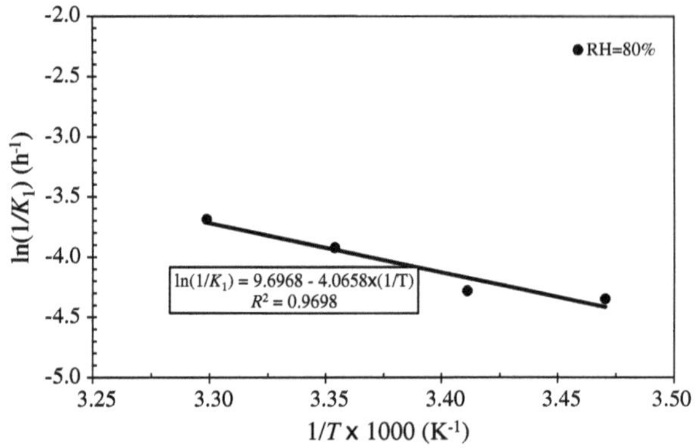


**Fig. 3.13** Effect of drying temperature on drying rate constant, for solid *red* brick samples at 25° C and RH = 80 %

## 3.5 Conclusions

Moisture is one of the most deteriorating factors of buildings. The moisture content is not only determined by the water that is absorbed by the material, but also by the amount of water that is evaporated under favourable conditions, which is described by the drying process.

The effect of drying temperature and relative humidity on drying of External Thermal Insulation Composite Systems (ETICS) and solid red brick samples were studied.

In order to explain the drying behavior of ETICS and solid red brick samples, different drying models, presented in the literature, were compared with their square regression co-efficients ($R^2$) and statistical parameters such as mean bias error (MBE), the root mean square error (RMSE) and chi-square ($\chi^2$). According to the results, change of moisture ratio with time can be described by Hii et al. and Midilli models in temperature range of 15–30° C and relative humidity range 50–80 %, for solid red brick samples. In case of the variation of these models constants or coefficients with temperature take into consideration. Hii et al. and Midilli models give the best result than the others models. The models developed have an excellent fit and would be a useful engineering application concerning ETICS and solid red brick samples.

In the future new tests with solid red brick samples must be carried out in order to confirm the results obtained. New materials will also be tested.

**Acknowledgments** J.M.P.Q. Delgado would like to thank Fundação para a Ciência e a Tecnologia (FCT) for financial support through the grant SFRH /BPD /84377 /2012.

## References

Barreira E, Freitas VP (2005) Importance of thermography in the study of ETICS finishing coatings degradation due to algae and mildew growth. In: Proceedings of the 10th DBMC international conference on durability of building materials and components, France

Bednar T (2002) Approximation of liquid moisture transport coefficient of porous building materials by suction and drying experiments. Demands on determination of drying curve. In: Proceedings of the 6th symposium on building physics in the nordic countries, Trondheim, Norway, p 493

Crank J (1975) The mathematics of diffusion. Oxford Science, New York

Delgado JMPQ, Ramos NMM, de Freitas VP (2006) Can moisture buffer performance be estimated from sorption kinetics? J Build Phys 29(4):281–299

Derdour L, Desmorieux H, Andrieu J (2000) A contribution to the characteristic drying curve concept: Application to the drying of plaster. Drying Technol 18(1):237–260

Krus M, Holm, A (1999) Simple methods to approximate the liquid transport coefficients describing the absorption and drying process In: Proceedings of the 5th symposium on building physics in the nordic countries. goteborg. Norway. p 241

Liu Q, Bakker-Arkema FW (1997) Stochastic modelling of grain drying: model development. J Agric Eng Res 66 :275–280

Maroulis ZB, Saravacos GD (2003) Food process design. Marcel Dekker, New York

Midilli A, Kucuk H, Yapar Z (2002) A new model for single-layer drying. Drying Technol. 20:1503–1513

Peleg M (1988) Empirical model for the description of moisture sorption curves. J Food Sci 53:1216–1219

White GM, Ross IJ, Ponelert R (1981) Fully exposed drying of popcorn. Trans ASAE 1981(24):466–468

Yagcıoglu A, Degirmencioglu A, Cagatay F (1999) Drying characteristic of laurel leaves under different conditions. In: Proceedings of the 7th international congress on agricultural mechanization and energy, May 26–27, Adana, Turkey

# Chapter 4
# Pore Structure Parameters and Drying Rates of Building Materials

**Peter Matiasovsky and Peter Mihalka**

**Abstract** Experimentally obtained courses of drying for representative porous building materials were investigated. With use of analytical and numerical solutions of the diffusion equation the moisture transport parameters of the materials were identified from the experimental data and modelled in relation to the parameters of their pore structure. The capillary models of moisture diffusivity based on the pore size distribution of given material were used in this analysis. It was confirmed that the character of drying rates specific both for their constant and falling periods is dependent on combination of only a few important pore structure parameters specific for each material. In concordance with the capillary models the pore volumes equivalent to critical degrees of water saturation during drying as the critical moisture content, the capillary moisture content and the full saturation are in a correlation with the drying rates and can serve as indicators of the actual evaporation potential of given material.

**Keywords** Drying process • Moisture diffusivity • Pore structure parameters • Modelling • Simulation • Building materials

## Nomenclature

| | |
|---|---|
| $A$ | Water absorption coefficient (kg/m$^2$ s$^{0.5}$) |
| $\beta$ | Surface water vapour diffusion coefficient (kg/m$^2$ s Pa) |
| $d$ | Thickness (m) |

P. Matiasovsky (✉) · P. Mihalka
Institute of Construction and Architecture, Slovak Academy of Sciences,
Dúbravská cesta 9, 845 03 Bratislava 45, Slovakia
e-mail: Peter.Matiasovsky@savba.sk

P. Mihalka
e-mail: Peter.Mihalka@savba.sk

J. M. P. Q. Delgado (ed.), *Drying and Wetting of Building Materials and Components*,
Building Pathology and Rehabilitation 4, DOI: 10.1007/978-3-319-04531-3_4,

| | |
|---|---|
| $D(w)$ | Material moisture diffusivity ($m^2/s$) |
| $D_a$ | Binary water vapour diffusion coefficient in air |
| $G$ | Relative evaporation (–) |
| $h_c$ | Convective surface heat transfer coefficient ($W/m^2$ K) |
| $K(w)$ | Permeability (s) |
| $p_{sat}$ | Saturation water vapour pressure (Pa) |
| $p_{crit}$ | Material water vapour pressure at critical moisture content (Pa) |
| $p$ | Water vapour pressure (Pa) |
| $p_c$ | Capillary pressure (Pa) |
| $q$ | Drying rate density ($kg/m^2$ s) |
| $r$ | Pore radius (m) |
| $R$ | Gas constant (J/kg K) |
| $S$ | Degree of saturation (–) |
| $SSA$ | Specific surface area ($m^2/m^3$) |
| $t$ | Time (s), water film thickness (m) |
| $T$ | Temperature (K) |
| $t_d$ | Duration of the first period drying (h) |
| $w$ | Volumetric moisture content ($kg/m^3$, $m^3/m^3$) |
| $w_{cap}$ | Capillary moisture content ($kg/m^3$, $m^3/m^3$) |
| $w_{crit}$ | Critical moisture content ($kg/m^3$, $m^3/m^3$) |
| $x$ | Space coordinate (m) |
| $\varphi$ | Relative humidity (%) |
| $\xi(\omega)$ | Moisture capacity (kg/kg Pa) |
| $\mu$ | Diffusion resistance factor (–) |
| $\rho$ | Density ($kg/m^3$) |
| $\eta$ | Dynamic viscosity (Pa s) |
| $\sigma_{lg}$ | Surface tension for water (N/m) |

## 4.1 Introduction

The transport parameters of inorganic porous building materials are given by their pore structure. Therefore besides the bulk density the material porosity serves in the first approach as the basic characteristics of material transport properties since the permeability of a material is proportional to its porosity in general, usually open in case of silicate based materials. However, the total open porosity not always represents the pore volume participating at transport process. At a specification of the porosity relevant to transport it is necessary to distinguish between the total material porosity and the porosity obtainable by a used pore structure parameters determination method, limited by the range of pores analysed, which is specific for given measurement technique or device. Term microporosity usually characterises the pore volume which is the most significant for the moisture transport. Some building materials e.g. the autoclaved aerated concrete are characteristic by the pore size distribution which has a multimodal character, given by the inclusions of large artificial spherical macropores dispersed in microporous matrix which do not participate on the moisture transport. In such cases it is distinctively important to distinguish between total porosity and microporosity.

Complex information on the material pore structure can be obtained by standard way with the use of mercury intrusion porosimetry (MIP) and the BET nitrogen or water adsorption and desorption tests. There is compatibility of the water sorption with results of the BET sorption and MIP techniques. Water vapour and nitrogen sorption experiments belong to the BET sorption techniques. The determination of pore size distribution by MIP is the intrusion technique. Concerning the wide range of analysed pores the pore structure parameters of building materials are mainly determined by the mercury intrusion porosimetry. The output of the MIP measurement is the pore size distribution of a material, which is considered equivalent to the water retention curve. The water retention and the mercury intrusion are complementary pores empting/filling processes characterised by the equivalence of Laplace and Waschburn equations during draining and intrusion respectively.

The water transport phenomena are usually modelled by the diffusion type equation. The equation can have two basic forms. In the first case the driving potential at moisture transport is represented by the capillary pressure and the transport parameters in the equation are moisture capacity and moisture permeability and the equation has so called capacity form. Parameters as the moisture permeability and capacity can be modelled from the pore size distribution or retention curves in this case. In the second case the water flow driving potential is represented by the moisture content and the material transport parameter is the moisture diffusivity and the equation has so called diffusion form (Bruce and Klute 1956). The moisture diffusivity can be determined from the moisture content profiles measured during transient water transport experiments. As the water transport parameters depend on the boundary conditions there is a difference between the moisture diffusivity during the water absorption, redistribution and drying and also different techniques are used at determination of moisture diffusivity. In a case of drying the diffusivity is smaller (Kuenzel 1995). A detailed study on drying diffusivity and its determination is given by (Pel et al. 1996, 2002; Pel and Landman 2004) which present the approaches to diffusivity determination at the absorption and at the drying during different drying phases. Vu (2006) realised the simulation experiments of drying process considering the variability of pore structure parameters expressed by theoretical pore size distributions and modelling the water transport coefficients of building materials represented as a bundle of capillaries, size distribution of which is based on their pore structure. In this chapter the principles of drying dynamics of the silicate based porous building materials are described and analysed in a connection with their pore structure parameters.

The drying experiments carried out for 25 building materials including burnt clay bricks, renovation plasters, mortars, calcium silicates, insulation plasters, lightweight carbonates and autoclaved aerated concretes are compared with simultaneously made MIP tests and the water absorption experiments. The obtained results were compared mutually and analysed with use of standard analytical and numerical solutions of moisture diffusion equation under boundary conditions at a drying.

## 4.2 Pore Structure Parameters

A standard method for determination of pore structure parameters is the mercury intrusion porosimetry. The method is based on the premise that a non-wetting liquid (one having a contact angle greater than 90°) will only intrude capillaries under pressure. The relationship between the pressure and capillary diameter is described by Washburn (1921).

For analysed building materials the relative pressure-volume data were determined by the 2002 mercury intrusion porosimeter and the micropore unit 120 of ERBA Science, enabling the determination of pores with radii from 3.7 nm up to 0.06 mm. The pore size distribution is determined from the volume intruded at each pressure increment. The total porosity is determined from the total volume intruded. The specific surface area of pores is calculated from the pore radii and the pore volumes. In Table 4.1 for example there is the pore size distribution given by MIP for one of the analysed materials, lime-cement insulation plaster.

As the unsaturated water transport represents two-phase flow the water transport parameters are the functions of the water saturation in a porous material (Bachmat and Bear 1991). The pore volume of the material from the aspect of a changing mechanism of fluid flow at different degrees of saturation is divided into three parts defined by two critical volumes necessary for percolation of the smallest and largest pores filled in saturation process which starts by filling the smallest pores. These critical pore volumes are given by the pore structure of given material and during the water saturation or drying process they are needed for a connected capillary pores network to be formed through the material, enabling water flow and air flow. The first of them, the critical moisture content is defined as the lowest moisture content necessary to initiate moisture transport in the liquid phase (Kumaran 1996). The second of them the critical pore volume, needed for a connected capillary pores network to be formed through the material, enabling air flow is a difference between saturation and capillary moisture contents due to entrapped air. The capillary moisture content is the value obtained in a short-term capillary rise experiment when the air initially present in the pores is partly entrapped preventing absorption until full saturation (Descamps 1997). The saturation moisture content corresponds to the full water saturation state.

The equivalence of the mercury intrusion and water retention curves means also the equivalence of critical percolation pore volumes determined by water saturation and mercury intrusion. The critical moisture content or irreducible wetting fluid saturation is equivalent to the volume of pores smaller than hydraulic radius. The results of analyses for building materials considered in this study

**Table 4.1** Pore structure parameters of lime-cement insulation plaster determined from mercury intrusion porosimetry

| Pore radii interval (nm) | Fraction pore volume ($m^3/m^3$) | Cumulative pore volume ($m^3/m^3$) | Note |
|---|---|---|---|
| 3.7–5.8 | 0.006 | 0.010 | (1) |
| 5.8–10 | 0.006 | 0.016 | |
| 10–18 | 0.024 | 0.040 | |
| 18–30 | 0.024 | 0.063 | |
| 30–58 | 0.035 | 0.099 | |
| 58–100 | 0.035 | 0.134 | (2) |
| 100–180 | 0.041 | 0.175 | |
| 180–300 | 0.047 | 0.222 | |
| 300–580 | 0.065 | 0.287 | |
| 580–1,000 | 0.112 | 0.399 | (3) |
| 1,000–1,800 | 0.100 | 0.500 | |
| 1,800–3,000 | 0.053 | 0.553 | |
| 3,000–5,800 | 0.012 | 0.565 | |
| 5,800–10,000 | 0.012 | 0.576 | |
| 10,000–18,000 | 0.018 | 0.594 | |

(1) Critical volume—volume of pores smaller than hydraulic radius/critical moisture content
(2) Volume relevant for capillary water transport
(3) Critical volume—volume of pores bigger than threshold radius/entrapped air at water absorption

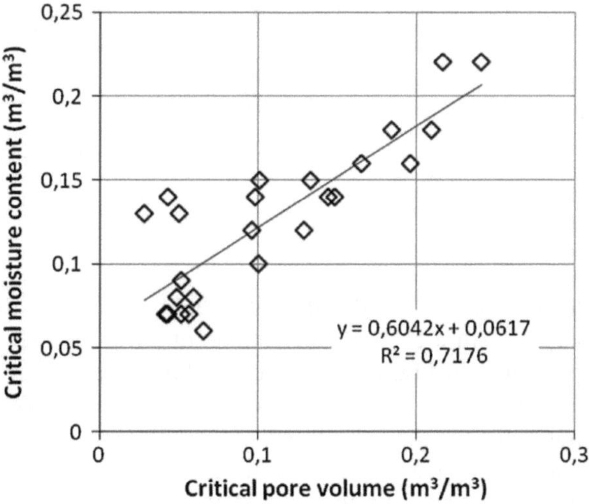


**Fig. 4.1** Correlation between critical pore volume (pores smaller than hydraulic radius) and critical moisture content

confirmed the correspondence between the critical moisture content and the pore volume smaller than hydraulic radius calculated from porosity and pore surface area resulting from MIP (Fig. 4.1). The capillary moisture content is equivalent

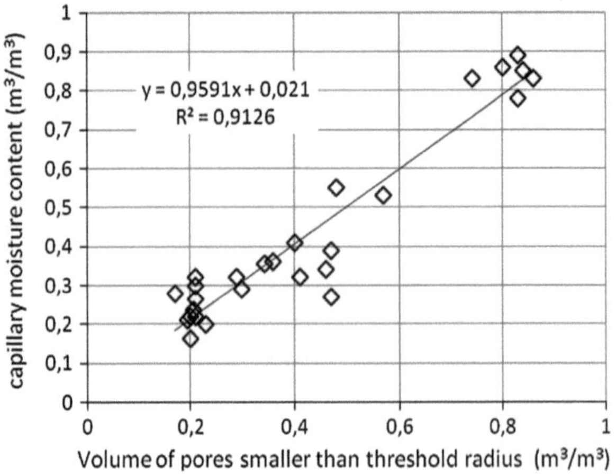


**Fig. 4.2** Correlation between volume of pores smaller than threshold radius and capillary moisture content

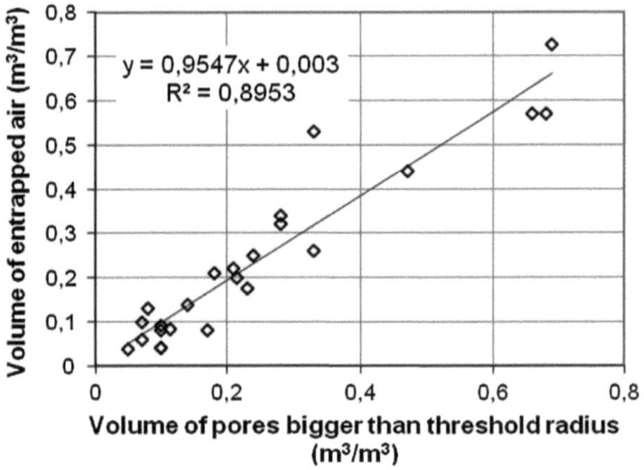


**Fig. 4.3** Correlation between volume of pores bigger than threshold radius and volume of entrapped air

to the volume smaller than threshold radius and the correlation between capillary moisture content and volume of pores smaller than MIP threshold radius is presented in Fig. 4.2. The pore volume critical for air outflow during the water absorption, or residual nonwetting fluid saturation, is equivalent to volume of pores higher than threshold radius. The value of critical volume for air transport (entrapped air) is close to the critical volume at the mercury intrusion process defined by the first inflection point of the cumulative intrusion curve when mercury connected pathway across the sample (threshold radius) (Hall and Hoff 2002). The comparison in Fig. 4.3 confirms this equivalence. The relevant porosity equals the total porosity reduced by the critical volumes. In Table 4.1 there are marked critical volumes including the volume relevant for liquid water transport and defined as the total porosity reduced by critical volumes.

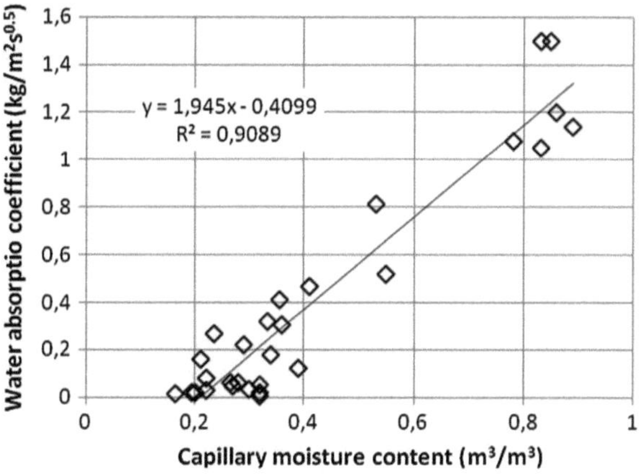


**Fig. 4.4** Correlation between capillary moisture content and water absorption coefficient

## 4.3 Moisture Transport in Building Materials and Their Pore Structure

The moisture transport in porous building materials for one dimensional problem is modelled by the diffusion equation in the form:

$$\frac{\partial w}{\partial t} = D(w) \cdot \frac{\partial^2 w}{\partial x^2} \tag{4.1}$$

where $D(w)$ is the moisture diffusivity which is a function of the moisture content. Besides an experimental determination of the moisture diffusivity there are the approaches trying to model this parameter on the pore structure basis, mainly the material capillary network models. These approaches issue from the capacity form of diffusion equation. Such an approach is applied for example at the modelling of water absorption coefficient where the model is based on the sharp moisture penetration profile (Hall and Hoff 2002) assumption:

$$A = \rho_w \cdot w_{cap} \cdot \sqrt{\frac{r_H \cdot \sigma_{lg}}{2 \cdot \eta}} \tag{4.2}$$

The model many times satisfactorily expresses the absorption coefficient and the dependence for the analysed materials in Fig. 4.4 confirms the significant correlation between capillary moisture content a water absorption coefficient.

Similarly also the moisture diffusivity can be modelled by the capacity form of diffusion equation as the diffusivity of network of serial or parallel capillary pore fractions (Descamps 1997) and the diffusivity is then expressed as the ratio of moisture permeability and moisture capacity:

$$D(w) = \frac{K(w)}{\xi(w)} \tag{4.3}$$

The diffusivity model (4.3) has three forms according to the moisture content region or degree of saturation. In the hygroscopic region up to ca 60 % relative humidity the water vapour diffusion is a dominant moisture transport mechanism and the moisture permeability of pore volume fraction has the form (Carmeliet et al. 2004):

$$K(w) = \frac{D_a}{\mu}(1-S)\frac{\varphi}{\rho_w R^2 T^2} \tag{4.4}$$

In Eq. (4.4) the determining material parameter is the diffusion resistance factor $\mu$.

Under the relative humidity higher than 60 % up to critical moisture content the simultaneous water vapour diffusion and surface flow are active (Deryaguin et al. 1985) and the moisture permeability of a pore volume fraction has the form:

$$K(w) = \frac{D_a}{\mu}(1-S)\frac{\varphi}{\rho_w R^2 T^2} + \frac{\Delta SSA \cdot t^3}{3 \cdot \eta} \tag{4.5}$$

The surface flow is determined by the specific surface are fraction $\Delta$ *SSA* and the water film thickness *t* at given moisture content.

If the moisture content is higher than critical the capillary flow is dominant and the moisture permeability of a pore volume fraction equals (Marshall 1958):

$$K(w) = \frac{\Delta w^2 \cdot r^2}{8 \cdot \eta} \tag{4.6}$$

The capillary flow is determined by the water filled fraction *w* and the pore radius corresponding to the pores saturation at given moisture content.

The moisture capacity is determined as the derivation of retention curve. The hygroscopic part BET water adsorption curve (Tables 4.2 and 4.3) applied in Eqs. (4.4) and (4.5) is connected with the over hygroscopic desorption branch modelled by MIP curve applied in Eq. (4.6). Vu (2006) analysed the drying process numerically for different materials with theoretical pores size distribution. In this study the pore size distributions of real materials are taken into consideration.

## 4.4 Drying Process

In comparison with the water absorption the drying is a slower process which reflects in smaller moisture diffusivities. During the drying in general two basic periods can be distinguished (Mujumdar and Devahastin 2000). The results for lime-cement perlite plaster will serve as an illustration.

1. The constant drying rate period. This first period is determined mainly by the boundary conditions which are given by the water evaporation from a wet material surface with constant moisture content and also by the material properties. The drying course is linear (Fig. 4.5) and the drying rate is constant (Fig. 4.6) during this period.

**Table 4.2** Water adsorption isotherm of lime-cement insulation plaster

| Relative humidity (−) | Equilibrium moisture content ($m^3/m^3$) |
|---|---|
| 0.12 | 5.84E-05 |
| 0.33 | 0.001694 |
| 0.53 | 0.003209 |
| 0.64 | 0.003853 |
| 0.75 | 0.004961 |
| 0.80 | 0.005116 |
| 0.85 | 0.006122 |
| 0.94 | 0.011533 |
| 0.98 | 0.022821 |

**Table 4.3** Basic material parameters of lime-cement insulation plaster

| Parameter | Value | Method |
|---|---|---|
| Porosity | 0.59 $m^3/m^3$ | MIP |
| Bulk density | 600 $kg/m^3$ | Gravimetric |
| Specific surface area | $10.47 \times 10^6$ $m^2/m^3$ | BET water adsorption |
| Hydraulic radius | $58 \times 10^{-9}$ m | MIP |
| Critical volume | 0.10 $m^3/m^3$ | Drying, MIP |
| Capillary moisture content | 0.356 $m^3/m^3$ | Free imbibition |
| Water absorption coefficient | 0.422 $kg/m^2\ s^{0.5}$ | Free imbibition |
| Diffusion resistance factor | 12 | Dry cup |

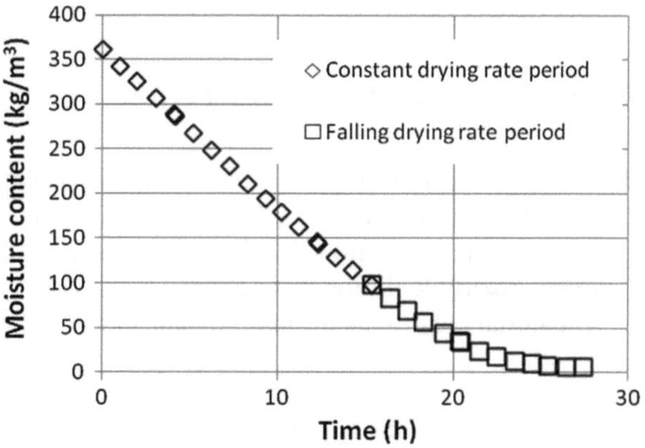


**Fig. 4.5** Measured time course of moisture content for lime-cement perlite plaster plate with differentiation of basic drying periods

2. The falling drying rate period. This second phase has in many cases the exponential time dependent course. The time course is determined by decreasing difference between material and surrounding water vapour pressures. The drying course is exponential (Fig. 4.5) and the drying rate is decreasing (Fig. 4.6), starting at the critical moisture content.

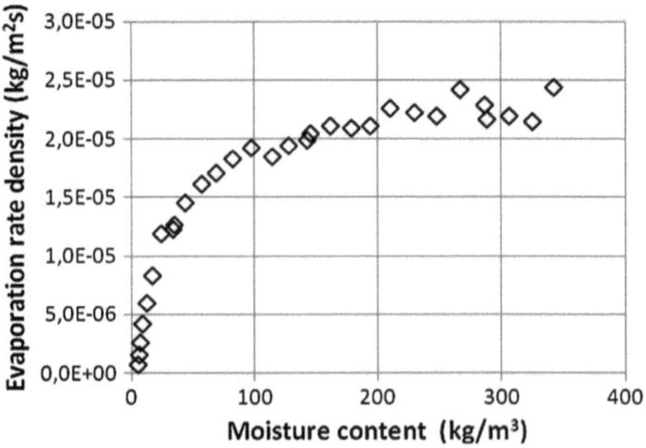


**Fig. 4.6** Variation of evaporation rate density with moisture content for lime-cement perlite plaster plate

During the experiments done for building materials the first drying period started at the capillary moisture content and it lasted until reaching the critical moisture content in the 50 % relative humidity and 20 °C temperature laboratory environment.

The second drying period started at the critical moisture content until reaching the moisture content equilibrium under the same constant laboratory boundary conditions as during previous period.

The drying experiments were carried out on the thin plate specimens of ca 0.01 m thickness, with 0.04 × 0.16 m dried surfaces exposed symmetrically at a vertical position. After processing of the obtained results the particular drying periods were analysed. The realised drying tests were considered as one dimensional moisture flows in infinite plates.

### 4.4.1 Constant Drying Rate Period

The drying course during this first period can be calculated as the solution of Eq. (4.1) under the boundary conditions of the second kind which are given by the constant water vapour flow density. The moisture diffusivity is so high that there is no moisture profile during this period.

The initial conditions for infinite plate of thickness $d$ are:

$$w(t = 0, 0 < x < d) = w_{cap} \tag{4.7a}$$

The boundary conditions of the second kind are:

$$D(w) \cdot \frac{\partial w}{\partial x} = q \quad (t > 0, x = 0, x = d) \tag{4.7b}$$

Under defined boundary conditions the analytical solution of Eq. (4.1) for average moisture content time course is (Carslaw and Jaeger 1959):

$$\bar{w}(t) = w_{cap} - \frac{q \cdot t}{(d/2)} \tag{4.8}$$

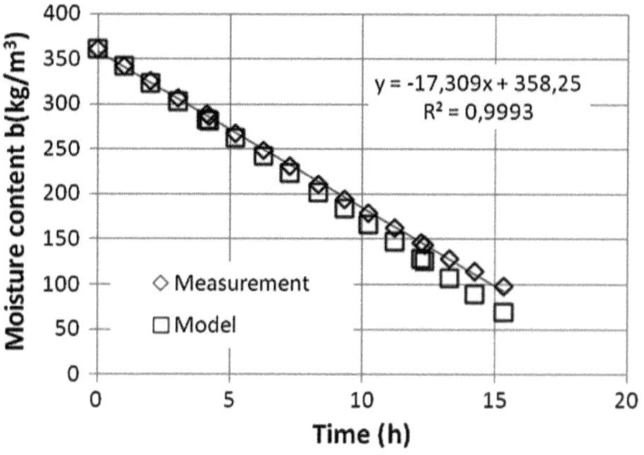


**Fig. 4.7** Time course of moisture content during constant drying rate period for lime-cement perlite plaster plate

In Fig. 4.7 there is the measured time course of moisture content during constant drying rate period for lime-cement perlite plaster compared with the analytical solution according to Eq. (4.8).

The drying rate is caused by the difference between capillary and critical moisture contents:

$$q = \frac{\left(w_{cap} - w_{crit}\right) \cdot (d/2)}{t_d} = G \cdot q_p \tag{4.9}$$

Term $G$ is the relative evaporation, the ratio of actual to potential evaporation (Granger and Gray 1989). The evaporation potential is calculated as:

$$q_p = \beta \cdot (p_{sat} - p) \tag{4.10}$$

where the surface water vapour diffusion coefficient is coupled with the convective surface heat transport coefficient (IEA Annex 14 1991):

$$\beta = 7.4 \cdot 10^{-9} h_c \tag{4.11}$$

The drying rate during the first period was analysed from the aspect of the actual to potential drying rate ratio $G$. For the unit surface element of porous medium the following approximation is valid:

$$q_p \approx G \cdot \beta \cdot (p_{crit} - p) \tag{4.12}$$

The interpretations of drying rate ratio concern the pore structure of drying materials and their surfaces. According to Or et al. (2007) and Shokri et al. (2010) during evaporation the liquid flows from the large capillaries to the evaporating surfaces of the small capillaries and the maximum capillary pressure difference for this capillary flow is given as:

$$\Delta p_c = 2\sigma \left( \frac{1}{r_{crit}} - \frac{1}{r_{cap}} \right) \tag{4.13}$$

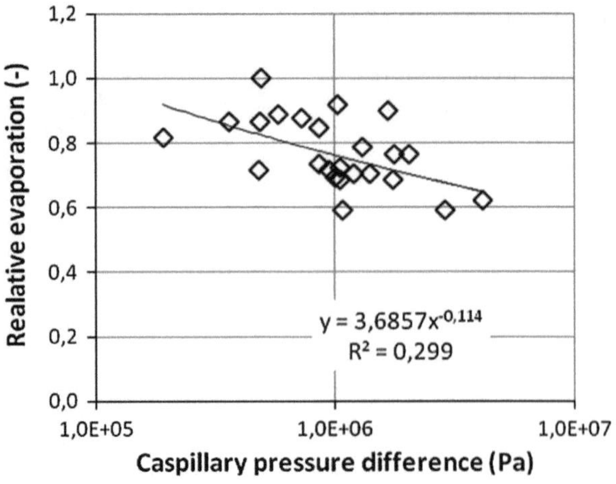


**Fig. 4.8** Dependence of relative evaporation on difference of capillary pressure at critical and capillary moisture contents

In Eq. (4.13) the largest pores correspond to the capillary moisture content and the smallest drainable pores contributing to continuous liquid phase conducting water to evaporating surface correspond to the critical moisture content. This capillary pressures difference is proportional to maximum drying front depth. The relationship between the capillary pressure difference determined from porosimetry and the relative evaporation determined from drying experiments for analysed building materials is in Fig. 4.8. The dependence of relative evaporation on the difference of capillary pressures expresses that the potential evaporation is reduced by the subsurface drying front. The minimum reduction is typical for the materials which have the biggest maximum radius of pores water filled with water at the critical moisture content.

Another factor determining the reduction of potential evaporation is that the drying surface becomes unsaturated in comparison with free water surface and the relative evaporation is caused by the actual surface—ambient water vapour pressure difference (Wilson et al. 1997). During the constant drying rate period the critical moisture content can considered on a surface.

Yiotis et al. (2003, 2004, 2005 and 2006) emphasized the role of surface film flow zone on drying and in a simple diffusion model the evaporation rate in a pore can be expressed as a proportional to the radius of water film. Assuming that the evaporation rate is proportional to the volume of pores filled with water one can state that the relative evaporation is proportional to the cube root of capillary moisture content. Combining the role of drying surface saturation and exposure of pores drained after decrease of capillary moisture content and covered by water film we can express the relative evaporation in form:

$$G \approx \sqrt[3]{w_{cap}} \cdot \frac{(p_{crit} - p)}{(p_{sat} - p)} \tag{4.14}$$

The relative evaporation values modelled according to Eq. (4.14) are compared with the measured data in Fig. 4.9 where evidently the relative evaporation is determined by capillary moisture content.

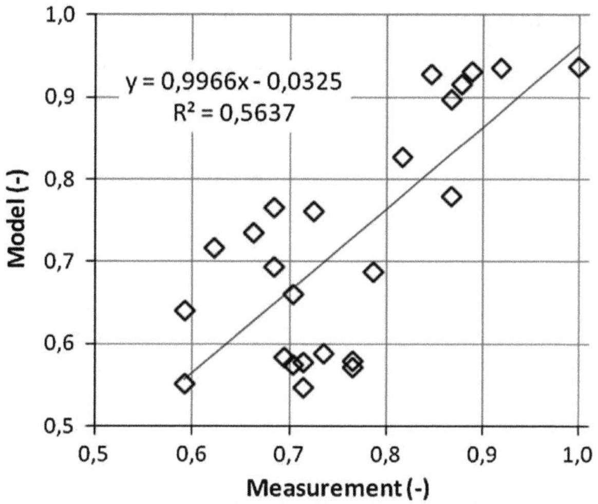


**Fig. 4.9** Comparison of relative evaporation measured and modelled according to Eq. (4.14)

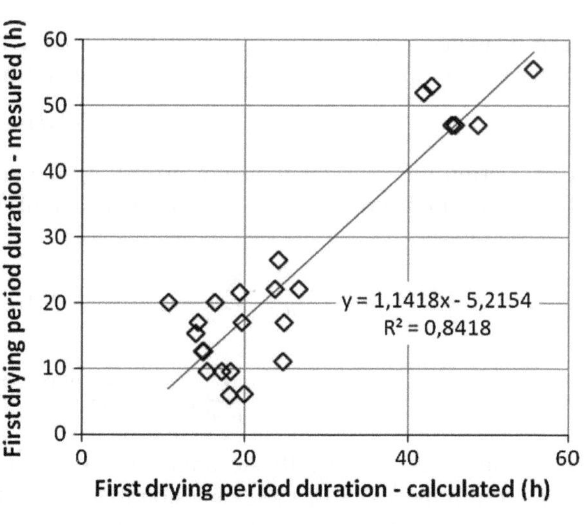


**Fig. 4.10** Comparison of measured and modelled values of constant drying period duration

With use of Eq. (4.8) the duration of the first drying period can be calculated. Applying the Eq. (4.14) this parameter was calculated and a comparison of the results with the results of drying experiments is in Fig. 4.10. A significant correlation is in agreement with the correlation between the capillary moisture content and the constant drying rate period duration which is described in Fig. 4.11.

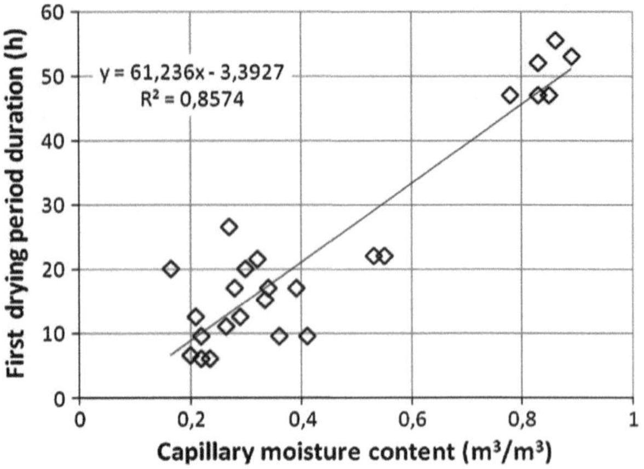


**Fig. 4.11** Dependence of first drying period duration on capillary moisture content

## *4.4.2 Falling Drying Rate Period*

During the falling drying rate period the solution of diffusion Eq. (4.1) for the infinite plate of thickness $d$ is calculated for initial conditions:

$$w(t = 0, 0 < x < d) = w_{crit} \tag{4.15a}$$

The boundary conditions of the first kind, considering that ambient and surface water vapour pressures are identical, are expressed as:

$$w(t > 0, x = 0, x = d) = w \tag{4.15b}$$

Considering the constant moisture diffusivity the analytical solution of Eq. (4.1) for average moisture content is calculated as (Crank 1975):

$$\begin{aligned}\bar{w}(t) &= w_{\infty} \cdot \frac{8}{\pi^2} \cdot \sum_{n=0}^{\infty} \frac{1}{(2n+1)^2} \exp\left(-\frac{(2n+1)^2 \pi^2 \bar{D}}{d^2} t\right) \\ &= w_{\infty} \cdot \frac{8}{\pi^2} \cdot \sum_{n=0}^{\infty} \frac{1}{(2n+1)^2} \exp(-m \cdot t)\end{aligned} \tag{4.16}$$

For long times, sufficiently accurate results are obtained by retaining only the first term in the series of the expression (4.16) and the moisture diffusivity can be expressed from measured drying course exponent $m$ as:

$$\bar{D} = \frac{m \cdot d^2}{\pi^2} \tag{4.17}$$

The exponent for lime-cement perlite plaster is determined by drying course curve fitting (Fig. 4.12). During the second period the drying represents the process which started at the critical moisture content. The time course of drying has an exponential character at all analysed materials, which confirms the validity of an assumption of constant diffusivity. From Eq. (4.16) was possible to determine the moisture diffusivity.

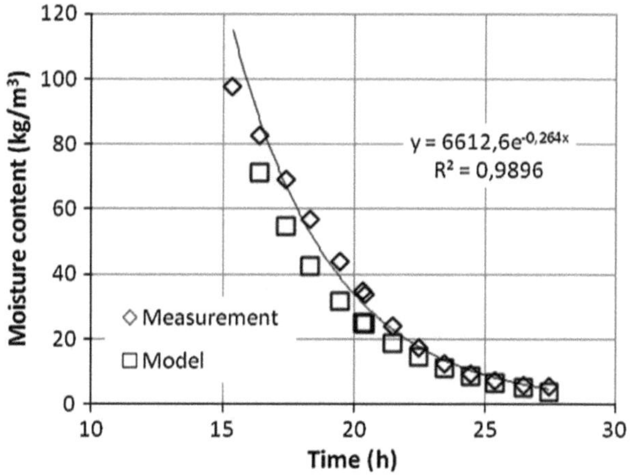


**Fig. 4.12** Time course of moisture content during falling drying rate period for lime-cement perlite plaster plate

In all experiments the constant moisture diffusivity is apparent in spite the fact that moisture diffusivity is the function of moisture content. However, various authors confirm a possibility of the constant moisture diffusivity (Mabirizi and Bulut 2009; Derdour and Desmorieux 2004) or the exponential character of drying (Mujumdar and Devahastin 2000; Barreira et al. 2013) during the second period. This can be explained by the fact that during the second drying period the drying rate is controlled by the constant average diffusivity of a specimen.

## 4.5 Numerical Simulation

In order to analyse the second drying rate period in detail the 0.008 m thick infinite lime-cement perlite plaster plate drying course was calculated with use of the numerical simulation program Delphin 4.2 (Grunewald 2000). The diffusion form of Eq. (4.1) was applied and the moisture diffusivity function was modelled in detail according to Eqs. (4.3–4.6). The diffusivity function is described in Fig. 4.13. The function at moisture contents lower than critical is characterised by its increase with decreasing moisture content and it was calculated from the permeability modelled by Eq. (4.5) using the water vapour adsorption data in Tables 4.2 and 4.3. The obtained diffusivity course shape is similar to the courses presented by (Pel et al. 1996, 2002; Pel and Landman 2004; Lockington et al. 2003; Li et al. 2008). The increase of the diffusivity function with decreasing moisture content is caused by the lower decrease of moisture permeability with steeper decrease of moisture capacity (Eq. 4.3) in the hygroscopic region. The maximum diffusivity in hygroscopic region was calculated according to Eq. (4.3). The diffusivity for moisture contents higher than critical was modelled according to Eq. (4.6) from the porosimetry data in Table 4.1.

The results of the simulation compared with experiment confirm the correctness of the diffusivity model used. Figure 4.14 shows a variation of the calculated evaporation rate density with moisture content in comparison with the measurement

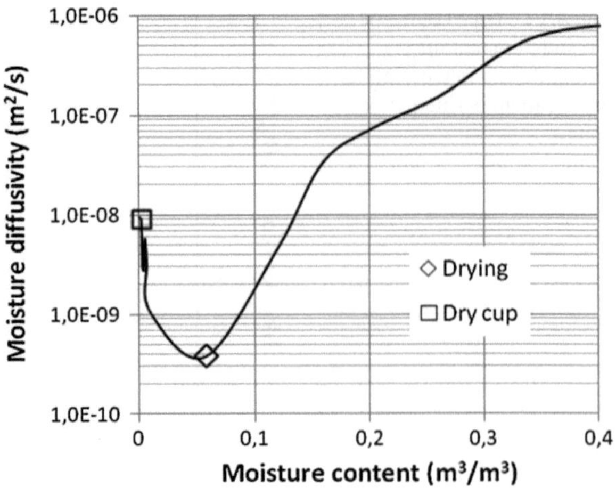


**Fig. 4.13** Moisture diffusivity of lime-cement perlite plaster with marked values determined from dry-cup test and drying experiment

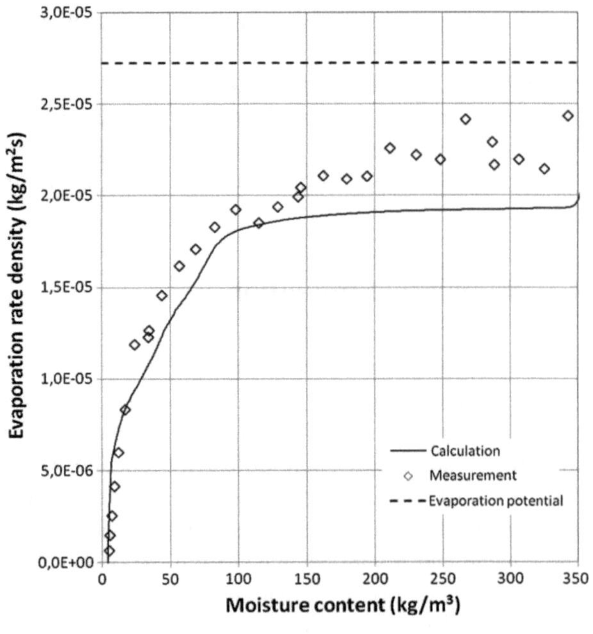


**Fig. 4.14** Variation of evaporation rate density with moisture content for lime-cement perlite plaster plate—comparison of calculation with measurement

when only the fluctuation of evaporation rate during the first period is visible. The surface water vapour diffusion coefficient value $2.96 \times 10^{-8}$ kg/m$^2$ s Pa was considered in the simulation. In Fig. 4.15 there is the calculated time course of moisture content in comparison with the measurement in a concordance of the calculation with the measurement, in spite of the discrepancy caused by instability of evaporation rate during the first period. The evolution of moisture content profiles in Fig. 4.16 was used for the analysis of moisture content dependence of moisture diffusivity during the falling drying rate period. The obtained results confirm the

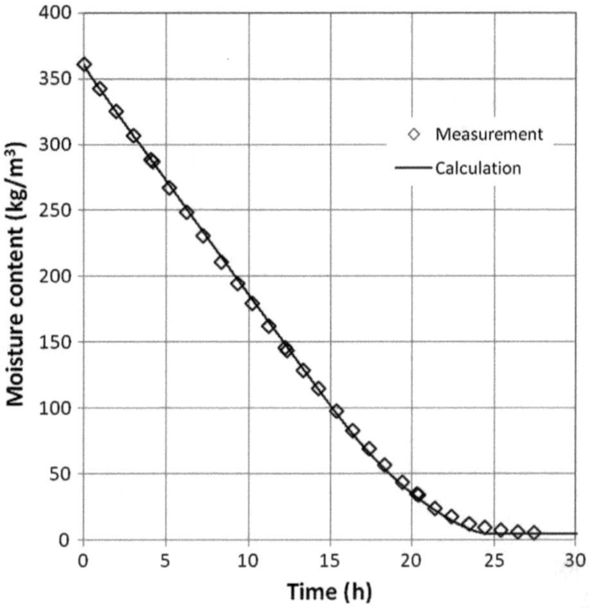


**Fig. 4.15** Measured time course of moisture content for lime-cement perlite plaster plate—comparison of calculation with measurement

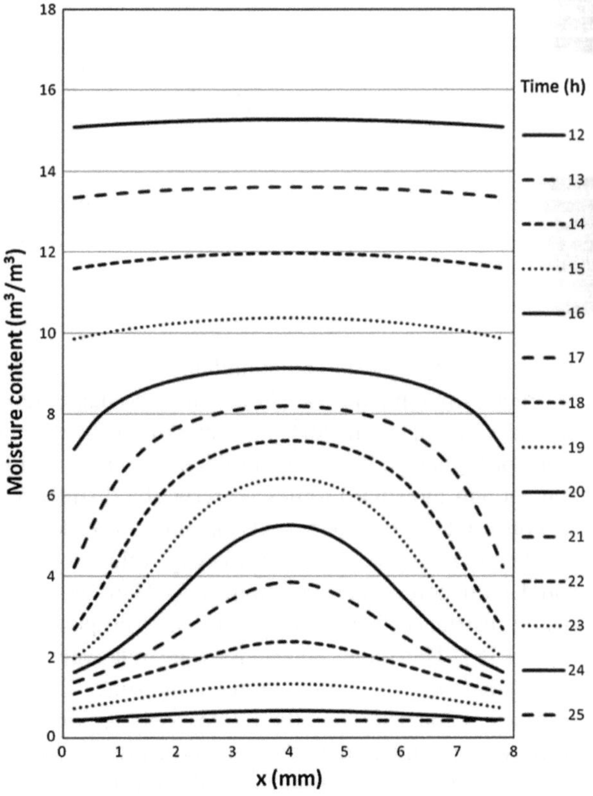


**Fig. 4.16** Calculated moisture content profiles in lime-cement perlite plaster plate during falling drying rate period

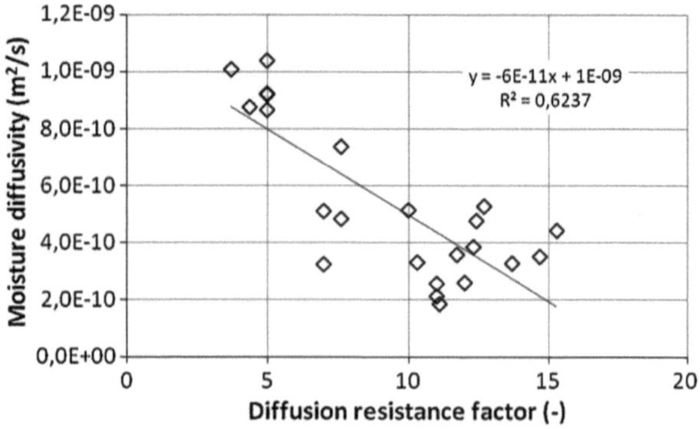


**Fig. 4.17** Correlation between diffusion resistance factor and moisture diffusivity during falling drying rate period

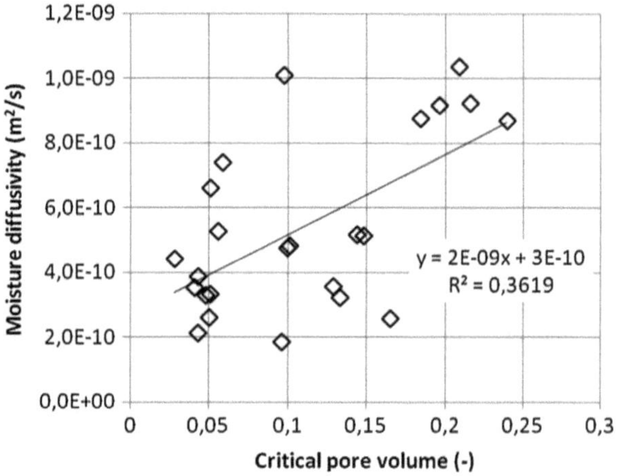


**Fig. 4.18** Correlation between critical pore volume and moisture diffusivity during falling drying rate period

creation of drying front and simultaneously the decrease of moisture diffusivity up to order of magnitude $10^{-10}$ m$^2$/s after he decrease of moisture content below the critical value of 0.1 m$^3$/m$^3$. The simulations of drying were repeated for the specimens of higher: 0.05 and 0.1 m thickness. The calculated time courses of moisture content and moisture profiles preserve the distinction between constant and falling -exponential drying rate periods also in these cases.

The minimum moisture diffusivity values obtained by Eq. (4.17) were compared with pore structure parameters. They are proportional inversely to the diffusion resistance factors (Fig. 4.17) and proportional to the critical pore volumes (Fig. 4.18). The found correlations are relatively weak, however they confirm the correspondence between minimum moisture diffusivity and some pore structure parameters characterising the hygroscopic region. A modelling of the moisture diffusivity function based on the pore structure parameters is a complex task and therefore it is difficult to determine diffusivity values by a single pore structure parameter.

## 4.6 Conclusions

The drying rate of porous building materials depends on their pore structure. The formulation of their transport parameters based on pore structure parameters gives the insight into this dependence. An analysis of the drying experiments carried out for various building materials in relation to their pore structure parameters obtained from mercury intrusion porosimetry tests confirmed that the drying rate is determined by two water saturation states: the critical moisture content and the capillary moisture content. As there is a straight relationship between the water saturation and mercury intrusion the critical moisture content corresponds to the volume of pores smaller than hydraulic radius and the capillary moisture content corresponds to the total open porosity reduced by volume of pores bigger than threshold radius. Therefore the porosimetry data can give equivalent information on the material pore structure from the aspect of drying. The first period of drying characterised by the constant drying rate is determined by two factors expressing the influence of pore structure on moisture transport in material and on boundaries. The first factor is the difference of capillary pressures between the initial—capillary moisture content and the final—critical moisture content during this period. The second factor is the exposure of capillary moisture content on the drying surface via the surface film flow expressed as the cube root of capillary moisture content, together with the difference of vapour pressures between the surface and ambient air. The second period of drying is characterised by the exponential drying time course indicating the constant average moisture diffusivity during this period. The values of this diffusivity are in the interval of $10^{-10}$–$10^{-9}$ $m^2/s$ for building materials. The detailed capillary model of moisture diffusivity function during drying shows that these values represent the minimum diffusivity at the moisture contents smaller than critical moisture content. Applying this modelled moisture diffusivity function in the numerical simulation of drying confirmed that during the second period the moisture diffusivity lies in the identified interval. The minimum moisture diffusivity is connected with surface flow and it is correlated with diffusion resistance factor and critical moisture content. The numerical simulations confirm the validity of these conclusions independently on the thickness of drying element.

## References

Barreira E, Delgado JMPQ, Ramos NMM, de Freitas VP (2013) Experimental evaluation of drying kinetics of building materials. In: 2nd central European symposium on building physics, Vienna, pp 745–750

Bear J, Bachmat Y (1991) Introduction to modelling of transport phenomena in porous media. Kluver Academic Publishers, New York

Bruce RR, Klute A (1956) The measurement of soil moisture diffusivity. Soil Sci Soc Am Proc 20:458–462

Carmeliet J, Hens H, Roels S, Adan O, Brocken H, Cerny R, Pavlik Z, Hal C, Kumaran K, Pel L (2004) Determination of the liquid water diffusivity from transient moisture transfer experiments. J Build Phys 27:277–305
Carslaw HS, Jaeger JC (1959) Conduction of heat in solids. Oxford at the Clarendon Press, Oxford
Crank J (1975) The mathematics of diffusion. Clarendon Press, Oxford
Derdour L, Desmorieux H (2004) Model for internal moisture diffusivity during the regular regime. Comparison with experimental data obtained on plaster and spirulina. In: Proceedings of the 14th international drying symposium, vol. A. São Paulo, pp 718–725
Descamps F (1997) Continuum and discrete modelling of isothermal water and air transfer in porous media. PhD dissertation, Leuven, KU
Deryaguin BV, Tshurayew NV, Muller VM (1985) Surface forces, Nauka. Moscow. (In Russian)
Granger RJ, Gray DM (1989) Evaporation from natural nonsaturated surfaces. J Hydrol 111:21–29
Grunewald J (2000) Documentation of the numerical simulation program DIM3. 1. Volume 2. User's guide. Insitute of building climatology, faculty of architecture. University of Technology Dresden, Dresden
Hall Ch, Hoff W (2002) Water transport in brick, stone and concrete. Taylor & Francis, New York
IEA-Annex 14 (1991) Condensation and energy. Final report, volume 1, source book
Kuenzel HM (1995) Simultaneous heat and moisture transport in building components. IRB Verlag, Stuttgart
Kumaran MK (1996) Heat, air, and moisture transfer in insulated enveloped parts. Final report. Task 3: material properties, IEA Annex 24, KU, Leuven
Li Ch, Li K, Chen Z (2008) Numerical analysis of moisture influential depth in concrete during drying-wetting cycles. Tsinghua Sci Technol 13:696–701
Lockington DA, Parlange J-Y, Barry DA, Leech CA (2003) Drying of porous building materials: hydraulic diffusivity and front propagation. Mater Struct 36:448–452
Mabirizi D, Bulut R (2009) Unsaturated soil moisture drying and wetting diffusion coefficient measurements in the laboratory. Report OTCREOS7.1-11-F, Oklahoma Transportation Center, Midwest City
Marshall TJ (1958) A relation between permeability and size distribution of pores. J Soil Sci 9:1–8
Mujumdar AS, Devahastin S (2000) Mujumdar's practical guide to industrial drying. Exergex Corporation, Watertown, Massachusetts
Or D, Lehmann P, Shokri N (2007) Characteristic lengths affecting evaporation from heterogeneous porous media with sharp textural boundaries. Estudios de la Zona No Saturada del Suelo vol. VIII. J.V. Giráldez Cervera y F.J. Jiménez Hornero, Cordoba
Pel L, Brocken H, Kopinga K (1996) Determination of moisture diffusivity in porous media using moisture concentration profiles. Int J Heat Mass Transf 39:1273–1280
Pel L, Landman KA, Kaasschieter EF (2002) Analytic solution for the non-linear drying problem. Int J Heat Mass Transf 45:3173–3180
Pel L, Landman KA (2004) A sharp drying front model. Drying Technol 22:637–647
Shokri N, Lehmann P, Or D (2010) Evaporation from layered porous media. J Geophys Res 115:B06204. doi:10.1029/2009JB006743
Vu TH (2006) Influence of pore size distribution on drying bahaviour of porous media by a vontinuous model. PhD thesis, Otto von Guericke University, Magdeburg
Washburn EW (1921) The dynamics of capillary flows. Phys Rev 17:273–283
Wilson GW, Fredlund DG, Barbour SL (1997) The effect of soil suction on evaporative fluxes from soil surfaces. Can Geotech J 34:145–155
Yiotis AG, Boudouvis AG, Stubos AK, Tsimpanogiannis IN, Yortsos YC (2003) Effect of liquid films on the isothermal drying of porous media. Phys Rev E 68:037303
Yiotis AG, Boudouvis AG, Stubos AK, Tsimpanogiannis IN, Yortsos YC (2004) Effect of liquid films on the drying of porous media. AIChE J 50:2721–2737
Yiotis AG, Stubos AK, Boudouvis AG (2005) Pore-network modeling of isothermal drying in porous media. Transp Porous Med 58:63–86
Yiotis AG, Tsimpanogiannis IN, Stubos AK, Yortsos YC (2006) Pore-network study of the characteristic periods in the drying of porous materials. J Colloid Interface Sci 297:738–748

# Chapter 5
# Setting and Drying of Bio-Based Building Materials

**Thibaut Colinart and Patrick Glouannec**

**Abstract** Bio-based materials become successful in the field of building construction because of their sustainability and environmental benefit. In the present chapter, we focus on one of them: Hemp Concrete. This material is made out of hemp shivs mixed with a pre-formulated lime based binder and water, and can be manufactured through three classical processes: spraying, moulding or mechanical mixing and tamping. Similarly to classical concrete, setting and drying are important stages since it influences the mechanical and hygrothermal properties. Here, setting and drying stages are investigated through different experiments performed on several instrumented blocks and on a large-scale wall during the curing time. The instrumentation (thermocouples, humidity sensors, weight-scales) allows investigating the drying kinetics and apprehending the hygrothermal behaviour of the material. It is observed that manufacturing process influences the initial water content whereas the hygrothermal behaviour depends on the material formulation. In addition, a model of multi-physics knowledge taking into account the various couplings is developed to predict desorption kinetics. Particular attention is paid in the experimental determination of hygrothermal properties. A roughly good agreement was found between simulations and experiments for different conditions.

**Keywords** Drying • Building materials • Bio-based materials • Hemp concrete • Hygrothermal properties

T. Colinart (✉) · P. Glouannec
EA 4250, LIMATB, Université de Bretagne Sud, 56100 Lorient, France
e-mail: thibaut.colinart@univ-ubs.fr

P. Glouannec
e-mail: patrick.glouannec@univ-ubs.fr

J. M. P. Q. Delgado (ed.), *Drying and Wetting of Building Materials and Components*, Building Pathology and Rehabilitation 4, DOI: 10.1007/978-3-319-04531-3_5, 

## 5.1 Introduction

Nowadays, building industry faces up to the sustainable development and the necessity of designing, building/renovating, operating and reusing buildings in an ecological and resource efficient manner (Green Building Home Page 2013). To achieve these goals, major actions must be addressed through the whole life-cycle of a building, such as energy saving methodologies and techniques, improved use of materials, their further reuse/recycle and emissions control. These points concerns in particular building material, which is one of the most important components of a sustainable building. Recently, Joseph and Tretsiakova-McNally (2010) focus on the recent advances in the field of the most common (cement/concrete; wood; brick; stone; ceramics; glass; plastics) and alternative (bamboo; cob; adobe) building material that have non-metallic nature. In the case of concrete and cement, they listed the main strategies dealing with challenges of modern cement and concrete industries. It includes:

- The improvement of durability, mechanical properties and service life of concrete, by developing and applying high performance concrete (HPC), self-compacting concrete (SCC), ultra-high performance cement composite or adequate reinforcing techniques.
- The reduction of the cement content in the concrete mix by increasing the application of supplementary cementitious materials, mainly pozzolanic or hydraulic materials (like fly ash, ground granulated blast furnace or silica fume).
- The increase in the use of recycled materials in place of natural non-renewable resources. It may include recycled concrete aggregate (RCA); crushed blast furnace slag; sand; brick; glass; granulated plastics; waste fibreglass; mineralized wood shavings; etc.

An alternative to the previous common building material consists in looking for new materials which significantly reduce resource and energy consumption (low embodied energy and minimal processing requirement), provide a better energy efficiency without causing pollution and damaging health and eco-systems (Woolley 2006; Glass et al. 2008). In this respect, bio-based (or green) materials originated from renewable organic constituents of plants and animals are good candidates for building construction. For example, agro by-products that mainly contain the stems of crop plants (like wheat, rice, sugarcane, etc.) or forestry wastes (like cork granules or bark of the trees) are mixed with a binder to form composites (Karade 2010).

Among the numerous lignocellulosic by-products, hemp shivs (or hemp hurds) which are the chopped remains of the woody stem of the plant are associated with a binder to form Hemp Concrete (HC). Originally, cement has been first used as binder (De Bruijn et al. 2009; Bütschi et al. 2004), but it was rapidly substituted by a pre-formulated lime-based mix of hydraulic and aerial lime and pozzolanic admixtures (Nguyen et al. 2009; Nozahic et al. 2012). With appropriate proportion of hemp and binder, HC can be manufactured with specific thermo-physical

**Table 5.1** Indicative thermophysical properties of hemp concrete depending on the type of application

| | Dry apparent density (kg $m^{-3}$) | Dry thermal conductivity (W $m^{-1}$ $K^{-1}$) | Elasticity modulus (MPa) | Compressive strength (MPa) |
|---|---|---|---|---|
| Wall | 400 | 0.1 | >15 | >0.2 |
| Roof | 200–250 | 0.06 | >3 | >0.05 |
| Floor | 500 | 0.12 | >15 | >0.3 |

Mechanical performances are required for samples stabilized at 20 °C and 50 % (Lanos et al. 2013)

criteria (apparent density, elasticity modulus, compressive strength, thermal conductivity, etc.) and thus cover various type of application in a building, like Roof insulation (minimal coating of hemp shivs to fix them to each other), Wall (good compromise between the thermal and mechanical properties) and Ground floor insulating slab (the more proportion of lime, the greater mechanical properties) (Lanos et al. 2013). Indicative values are summarized in Table 5.1.

Nowadays, the manufacturing methods for mixing and implementing HC correspond either to traditional techniques applied to common concrete (mixing in a cement or concrete mixer and casting in a formwork) or techniques based on the adaptation of more modern procedures (moulding or vibro-compacting of prefabricated blocks and on-site assembling, in situ spraying, etc.). Depending on the material formulation, the mixing and implementation processes, water is added to hemp and binder to form a fresh paste with a sufficient fluidity and to ensure an easy manufacturing (Lanos et al. 2013). Nevertheless, adjusting its amount is not simple work since water plays different roles during the setting and the curing of HC:

- Water may be trapped quickly by the hemp shivs during the mixing. Chamoin et al. (2011) and Nguyen et al. (2009) observed that hemp shivs can absorb up to 300 % in mass of water in less than 5 min because of their high porosity, small pore size and hydrophilicity. It means therefore that hemp shivs are somewhat in competition with the binder to mobilize water, but they can also release water if necessary during the curing. Currently, two ways are evaluated to improve the water management: ones look for the best granular size distribution to set HC (Nguyen et al. 2009); the second suggests treating hemp shivs with hydrophobing solution (Chamoin et al. 2011).
- Water interacts with every constituent of the pre-formulated lime-based binder. For example, the setting of hydraulic lime takes place in two stages. First, dicalcium silicate C2S (as well as the other minor phases C3S, $C_4AF$, $C_2AS$ and $C_3A$) are hydrated to form calcium silicate hydrate (C–S–H) and portlandite according to (Lanas et al. 2004):

$$2CaO \cdot SiO_2 + H_2O \rightarrow C - S - H + Ca(OH)_2 \quad (5.1)$$

This hydration reaction kinetic is relatively slow (few days) and is obviously enhanced under moist conditions. Secondly, the previously formed portlandite and the bulk calcium hydroxide $Ca(OH)_2$ contained in the aerial lime is carbonated on contact with the atmospheric $CO_2$ and free water is formed according to (Lawrence et al. 2007):

$$Ca(OH)_2 + CO_2 \rightarrow CaCO_3 + H_2O \tag{5.2}$$

This carbonation reaction occurs over several years and is promoted under humid conditions or in exposition to high level of $CO_2$. Since both reactions are slow and may delay the setting of sufficient mechanical properties at early age, pozzolanic admixture with higher reactivity are general added into the pre-formulated lime-based binder (Gu and Song 2010).

Recently, a major advance in the understanding of the two above phenomena was proposed by Faure et al. (2012). With the help of NMR, they could follow the evolution in time of the respective amounts of water in the different phases (binder paste and hemp shivs) within HC. It is shown that for a sealed "mixed and cast" sample a significant amount of water is initially absorbed in the hemp and then progressively migrates towards the binder, which supplies additional water for the hydration reactions, even after 3 months. For comparison, a sample subjected to drying under natural convection in room conditions (21 °C and 54 %rh) was also followed: the same process occurs, but at the time a fraction of water also migrates through capillary effects from the centre to the surface through a diffusion pathway and evaporates at the surface. Furthermore, hemp shivs dry faster than the binder, and thus, if drying does not occur too rapidly, some of the water initially in the hemp shivs transfer towards the binder possibly providing a significant additional amount of water available for the hydration process. Last, their sample was considered as dried after 3 months.

Finally, in case of convective drying, the setting of HC results from a competition between the water absorption within the hemp shivs, the setting reactions occurring within the lime-paste and the significant loss of mass on drying. This last stage is well theoretically and experimentally investigated in the case of common concrete (Thiery et al. 2007), but less studied in the case of HC. Up to now, drying experiments on HC have been carried only by Bütschi et al. (2004) and (Evrard 2008), who studied the effect of the material formulation (in terms of water to binder mass ratio) or the mixing process. However, these results were not interpreted in terms of evaporation rate and of transport parameters. On the other hand, some works are devoted to the determination of hygric properties, like sorption isotherms, water vapour permeability, liquid absorption coefficient or moisture buffer capacity (Evrard 2008; Collet et al. 2013; De Bruijn 2012), but are not used for investigating numerically the drying stage.

In this chapter, a specific experimental set-up and methodology is developed in order to investigate the drying of different samples at two different scales, namely the material scale and the walls scale. The objective is to complete the previous set of results and to explore experimentally the significance of the setting process, of the initial composition, of the investigated scale and of the operating conditions on the drying of HC, but also on the hydric transfer within the material. Furthermore, a simplified numerical model for the prediction of drying kinetic is presented.

**Table 5.2** Overview of the sample's formulation and preparation

| Setting process | Material reference | Hemp shivs | Binder paste or {Binder + Water} |
|---|---|---|---|
| Process A | Wall | 23 %wt | Formulated binder A1: 40 %wt<br>Water: 37 %wt |
| Process A | Floor 1 | 20 %wt | Formulated binder A2: 47 %wt<br>Water: 33 %wt |
| Process A | Floor 2 | 19 %wt | Formulated binder A3: 48 %wt<br>Water: 33 %wt |
| Process B | Wall | 16 %wt (with fibers) | Tradical PF 70: 36 %wt<br>Water: 48 %wt |
| Process C1 | Wall | 17 %wt | Tradical PF70: 35 %wt<br>Water: 48 %wt |
| Process C2 | Wall | 20 %wt | Formulated binder: 45 %wt<br>Water: 35 %wt |

Then a comparison between experimental data and numerical results is proposed by paying attention in the evaluation of thermal and hygric properties.

In this view, the chapter is divided as follow: Sect. 5.2 contains the samples preparation, the measurement methodology and experimental results whereas Sect. 5.3 presents the material characterization, the developed drying model and the comparison between experimental and simulated results.

## 5.2 Multi-Scale Experimental Studies

Numerous HC samples with different formulations were manufactured by industrial partners through three common processes. The manufacturing processes include: moulding and vibro-compacting of prefabricated blocks (*Process A*), mechanical mixing, casting in a formwork and tamping (*Process B*) and spraying (*Process C*). More details on these manufacturing processes are available in (Colinart et al. 2012). The different formulations include the wall formulation and the floor formulation. The mass proportion of binder and hemp shivs respects the reference formulation suggested by Lanos et al. (2013) for usual application. All sample formulation and preparation are summarized in Table 5.2.

### *5.2.1 Drying Experiments at the Material Scale*

At the material scale, drying stage of HC was investigated on numerous samples with dimensions of $30 \times 30 \times 16$ cm$^3$. The sample thickness was set to 16 cm since it corresponds to the half thickness of a usual HC wall. The two other dimensions are chosen to be approximately representative of commercial blocks.

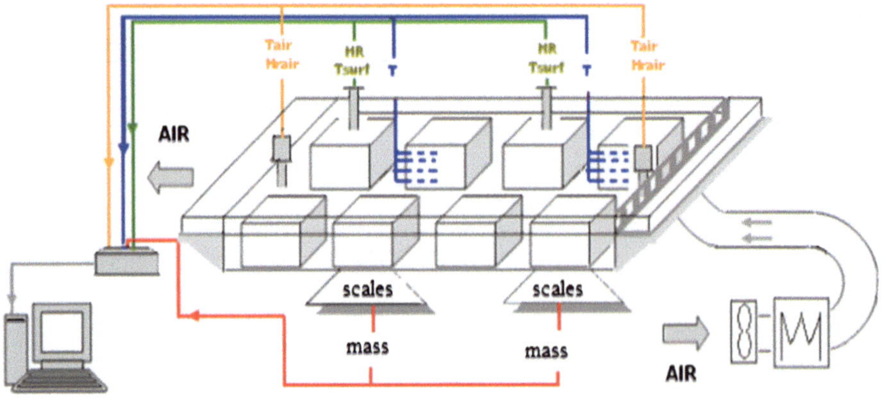


**Fig. 5.1** Schematic view of the air flue

After their fabrication by the industrial partners, the blocks were completely sealed during 2–5 days, in order to ensure a sufficient strength development of the material through the hydration reaction without exchanging water with the environment (Faure et al. 2012).

Once received in the laboratory, the drying stage of HC is performed either under natural convection within a climatic room or under forced convection in a specially designed air flue (see Fig. 5.1). In the context of a one-dimension thermal and mass transfer, only the upper face of the blocks is in contact with air. The drying kinetic is then investigated by measuring the product mass $m$ as a function of time with an electronic scale. In addition, numerous sensors are used to measure the temperature $T$ and relative humidity $\varphi$ variations at different locations within the blocks, as well in the air. All parameters (temperature, relative humidity, weight, air velocity) are recorded every 10 min through an acquisition system. Further information on the experimental device and the sensor positioning and accuracy can be found in (Colinart et al. 2012).

Drying experiments are performed until the product mass and the in situ relative humidity is stabilized. According to the norm EN 772-13 (2000), constant mass is reached, if during the drying process in two subsequent weighing with a 24 h interval, the loss in mass between the two determinations is not more than 0.2 % of the total mass. At the end of the drying experiment, the sample is dried in a ventilated oven at 60 °C until the complete dewatering of the product in order to get the dry mass $m_{dry}$, and thus the water content $w$ defined as:

$$w = \frac{m - m_{dry}}{m_{dry}} \tag{5.3}$$

It should be noted that mass measurement can be affected by the carbonation reaction, since molar mass of lime stone is 35 % higher than hydrated lime (Lawrence et al. 2007). However, this reaction is very slow and mass increase occurs only in several months. Thus, it can be supposed that the mass change

**Fig. 5.2** Wall projection and visualization of the wood-stud frame and the sensors

during the drying time can be attributed only to water loss of the material and that the dry mass is the same through the complete experiment.

Last, the drying stage is evaluated through two curves derived from the measurement of the sample's mass:

- The drying curves, based on the variation of the water content *w* in the product as a function of time,
- The drying rate curve (or Krischer's curve), defined as the speed of mass loss per drying surface unit as a function of time.

### 5.2.2 *Drying Experiments at the Wall Scale*

In addition to the blocks, a large-scale wall with dimensions of $270 \times 210 \times 36$ cm$^3$ (length × height × thickness) was sprayed (*Process C1* in Table 5.1) around a wood-stud frame in order to correspond to building standards (see Fig. 5.2). This wall was manufactured and dried within a bi-climatic chamber,[1] for which the chambers are regulated and monitored in temperature and relative humidity independently of each other.

Obviously, the experimental device allows not measuring the wall mass and water content. Nevertheless, a block was sprayed at the same time as the wall and dried under natural convection to access the initial water content within the

[1] www.limatb.univ-ubs.fr.

wall. On the other hand, a specific monitoring of temperature and relative humidity is performed with numerous K thermocouples and capacitive humidity sensors (type SHT 75 from Sensirion) at three main positions within the wall: at x = 7 cm (referred as "Interior"), at x = 18 cm (referred as "Middle") and at x = 31 cm (referred as "Exterior"). Additional instrumentation (optical pyrometer, thermohygrometers or hot-wire anemometer) allows apprehending the boundary conditions. More details on the experimental set-up can be found in (Colinart et al. 2013). Finally, the bi-climatic room test facility allows to:

- Evaluate temperature and relative humidity in time and in space,
- Examine the drying stage of sprayed HC and of plasters,
- Characterize the hygrothermal behaviour of HC as function of well defined climatic boundary conditions,
- Investigate the influence of plasters on the hygrothermal performances of the wall.

### 5.2.3 *Results*

Because of the specific interaction between hemp shivs and pre-formulated lime-based binder, drying experiments of HC should be performed under an optimal air relative humidity: if the curing conditions are too dry, carbonation occurs mainly at the block surface (skin effect), and thus reduced the pore size and the mass transport properties (Lawrence et al. 2007), which prevails any hardening in the centre of the material (powder aspect) (Cerezo 2005). If the curing conditions are too wet, the drying time can be dramatically increased. A good compromise lies in setting the relative humidity to 45–65 %rh at ambient temperature, i.e. 20–25 °C. Furthermore, these conditions are consistent with the results of Arnaud and Gourlay (2012), who found that a cure with 20 °C and 50 %rh lead to the highest compressive strength for their material.

#### 5.2.3.1 Blocks Drying

This part presents a typical result for HC block drying. Here, the drying under forced convection conditions of a sample manufactured according the *Process A* is investigated. Before analysing the experimental data, it should be noted that the experiment time is long (more than 4 months) and that the data acquisition was interrupted only 3 times (at the 15th, 30th and 110th day) due to power outage.

Figure 5.3a shows the evolution of temperature and relative humidity in the air flue. Air temperature is controlled at $T = 22$ °C during the drying period whereas the relative humidity oscillates between 30 and 60 %rh around an average value of 45 %rh. Figure 5.3b, c presents respectively the mean water content and the drying rate $F_m$ through the upper face of the sample as function of time. Both figures contains

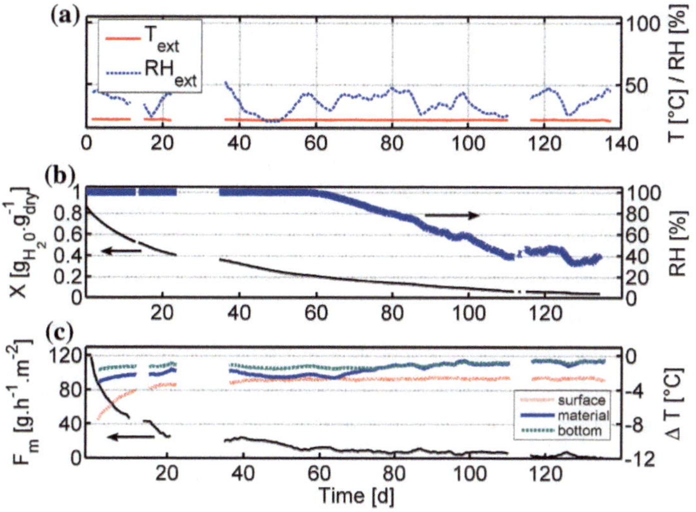


**Fig. 5.3** Drying under forced convection of HC manufactured according to the process A. Time evolution of experimental drying conditions (**a**), of mean moisture content (*black line*) and of relative humidity within the material (*bold blue line*) (**b**), of drying rate (*black line*) and of temperature difference between the sample and the air (**c**)

also the local measurements, i.e. the relative humidity measured at 8 cm from the upper face (see Fig. 5.3b) and the temperature differences between the air flue and respectively the surface, the middle and the bottom of the material (see Fig. 5.3c).

Initially, mean water content is about 0.87 g $g^{-1}$ and decreases immediately and continuously once drying begins. During the 25 first days, loss of water is important, as indicated by drying rate values ranging from 120 to 40 g $h^{-1}$ $m^{-2}$. During this period, moisture migrates from the material towards the surface and evaporates there. Since evaporation is an endothermic reaction, material surface temperature is about 8 to 4 °C lower than the air. Temperature differences observed within the material are also lower and can be due to heat conduction from the bottom to the surface, and not to evaporation, since relative humidity within the material is still at saturation.

Continuation of drying leads to slow decrease of material water content. Evaporation is progressively completed deep inside the material, as indicated by the decrease in the local the relative humidity measurement from the 60th day. Simultaneously, the temperature decrease observed in the middle of material between the 40th and 65th day (and not in the two other location) confirms that the evaporation front reaches this region. Once evaporated, moisture is then transported by diffusion to the surface at drying rate lower than 20 g $h^{-1}$ $m^{-2}$.

After, temperature increase again within the material and local relative humidity decrease progressively until the drying is completed. Looking to the mass variation, drying seems to be completed after 80 to 100 days. However, at this time, local relative humidity is not yet equilibrated with the environment. This is done

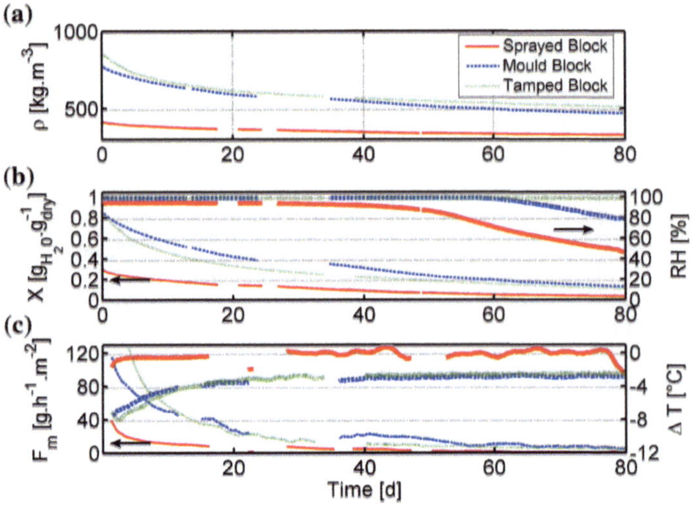


**Fig. 5.4** Influence of the setting process on the density (**a**), on the moisture content and on the relative humidity within the material (*bold lines*) (**b**), and on the drying rate and on the temperature difference between the sample and the air (*bold lines*) (**c**). Sprayed block (resp. mould block and tamped block) is represented by *solid line* (resp. *dashed* and *dotted line*)

around the 110th day: the relative humidity reaches values of about 40 %rh (which are close to those in the air), whereas temperatures are stabilized within the material. At the end of the process, the final water content is about 0.05 g $g^{-1}$, which is consistent with the one expected from the isothermal desorption curve (De Bruijn 2012). Finally, the material has a final density of about 420 kg $m^{-3}$ and an open porosity comprised between 0.5 and 0.6 (Glouannec et al. 2011).

The influence of the drying mode (natural or forced convection) and of the initial formulation was also investigated in Colinart et al. (2012). Results are not presented here, but it was shown that both points seem not to have an influence on the drying stage. Particularly, no major differences are observed between the two drying mode, since the air velocity in the air flue is low (less than 2 m $s^{-1}$). So, it was concluded that storage under ambient conditions could be sufficient to cure every kind of HC blocks, since they are produced on an industrial site. Furthermore, drying time could be reduced by subjecting all the faces to natural convection rather one.

#### 5.2.3.2 Influence of the Manufacturing Process on Blocks Drying

This part is devoted to the comparison between the different manufacturing *Processes A, B* and *C2*. In this view, Fig. 5.4a presents the density and Fig. 5.4b the local relative humidity measured within the material and the mean water content as function of time whereas Fig. 5.4c shows the drying rate $F_m$ and the temperature differences between the air flue and the surface of the samples as function of time.

Since their manufacturing processes are basically similar (step of mixing of ingredient and step for avoiding air bubble), the drying stage of mixed and tamped and the mould blocks are analogue: high initial density and water content, high drying rate and temperature difference between the air flue and the surface of the blocks at the beginning of the drying stage. As the drying proceeds, the mean water content decrease and the evaporation front moves slowly within the material. However, the local relative humidity is still at saturation level for long time and consequently, the drying time exceed 80 days for this kind of HC.

On the other hand, the sprayed blocks have lower initial water content (about 0.3–0.4 g $g^{-1}$) that the mould or mixed and tamped one (about 0.85 g $g^{-1}$), as it was expected. Consequently, the drying rate at the beginning of drying, as well the temperature difference between the air and the sample's surface are also lower. Thus, the evaporation rate at the surface is lower and the evaporation front migrates toward the centre of the material. This is confirmed by the fact that the relative humidity within the sprayed sample decreases only after 50 days. The equilibrium with the environment is reached in less than 70 days, which is faster than for the mould and mixed and tamped blocks. This point is advantageous in the view of applying wall render. Moreover, spraying process leads to low density materials, which have a lower thermal conductivity (Cerezo 2005; EN 12667 2001).

#### 5.2.3.3 Wall Drying

At the block scale, spraying show interesting advantages in terms of drying time. The objective of this part is to evaluate the drying stage at the wall scale. In this view, temperature and relative humidity distribution in the air and at different locations within the wall are plotted on Fig. 5.5.

At the beginning, the boundary conditions were not controlled since the doors of the climatic room were opened to remove a maximum of moisture after wall manufacturing. Then, the set points were fixed at T = 23 °C and $\varphi$ = 50%rh in each room of the bi-climatic chamber. Both graphs of Fig. 5.5 show that air temperature and relative humidity is accurately regulated in each room.

Looking to the temperature within the wall, we note that the three temperatures within the wall are approximately equal and follow the evolution of the interior and the exterior temperature. A small difference exists over the entire drying stage, but is reduced as drying proceeds. Evaporation effects may maintain temperature within the wall at lower level than in the air. Concerning the relative humidity, only one sensor was in use during the 100 first days, the two other being damaged due to the high moisture level within the wall. All thermohygrometers were replaced after 120 days. After 50 days, the evaporation front reaches the sensor located at 7 cm from the interior chamber. As a reminder, the evaporation front reached the sensor located at 8 cm from the top surface after 50 days (see Fig. 5.4). This difference may be explained by the difference in the initial water content: block initial water content is about 0.35 g $g^{-1}$, whereas wall initial water content is about 0.4 g $g^{-1}$. Such a difference can be explained by the fact that the

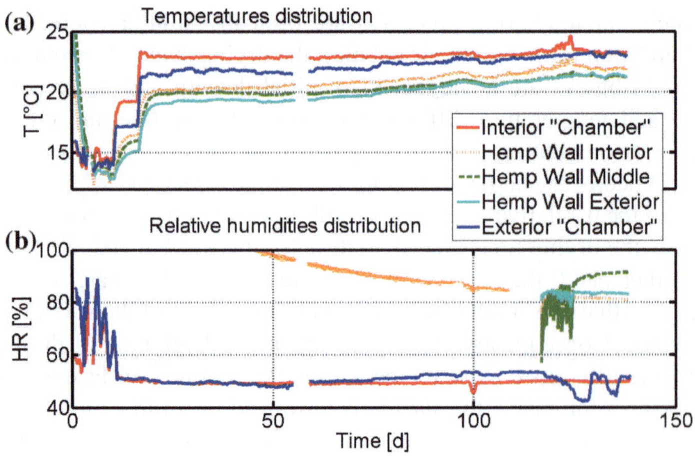


**Fig. 5.5** Drying of HC wall manufactured according to the Process C1. Time evolution of temperature (**a**) and relative humidity (**b**). Both graphs include the measurement performed in each climatic room and at different location within the wall

operator adjusts manually the amount of water during the HC spraying and may be thus responsible for the difference observed between the blocks and the wall. After the 50th day, relative humidity within the material decreases slowly. After 120 days, relative humidity measured close to the exterior room is in the order of magnitude than the one measured close to the interior room because of symmetry conditions, whereas the relative humidity measured at the wall centre is about 5 to 10 %rh higher. Even after 150 days, the relative humidity within the wall was not stabilized, indicating that the drying stage is probably not complete.

## 5.3 Numerical Simulation of Hemp Concrete Drying Stage

In this section, a simplified mathematical model is presented in the view of catching the drying kinetic of HC. Before developing the constitutive equations for heat and mass transfer (Sect. 3.2), attention is first paid in the determination of hygrothermal storage and transfer material properties in order to guarantee the accuracy of the input parameters for the simulations.

### *5.3.1 Material Properties*

#### 5.3.1.1 Thermal Properties

The thermal capacity $c_P$ is determined by differential scanning calorimetric measurement (DSC) that were carried out with a high-sensitivity SETARAM Micro DSC III micro-calorimeter. Measurements were made on 400 mg samples at 0.2 K min$^{-1}$

scan rate in the temperature range −5 to 40 °C. At least two measurements were performed on dry and wet hemp shivs, lime based binder and HC, with a view of validating mixing law for multi-component materials.

Thermal conductivity of building materials can be measured by different methods that are broadly classified as steady-state methods and transient methods. On the one hand, measurements were performed with a homemade guarded hot plate set-up according to the standard protocol defined in EN ISO 12667 (2001). Cold and hot temperatures are respectively set to 15 and 25°C and the set-up is placed in a climatic room (20 °C, 50 %rh). Experiments were performed either on dry or room-conditioned samples with dimensions of $100 \times 100 \times 30\ mm^3$. However, steady-state methods suffer from major drawbacks (long experimental time, large temperature gradient, large sample size, etc.). Therefore, thermal conductivity was also measured using the hot-strip transient methods. Details on the experimental protocol and measurement interpretation can be found in Pierre et al. (2013). One of the main advantages of this technique is the short measurement time, and thus evaporation/condensation phenomena are limited during the measurement on moist samples. Consequently, numerous thermal tests were carried out on sprayed HC under controlled conditions in temperature (from −3 °C to 30 °C ± 1 °C) and in relative humidity (from 0 to 95 %rh). Results indicate that thermal conductivity increases with temperature and relative humidity, and thus with water content. Furthermore, experiments performed on dry samples shows that there are no significant differences between the hot-strip and the guarded hot plate measurements.

For hygrothermal simulations, a constant value of 1.07 J/g/K (measured at 20 °C on a dry material) may be chosen for the thermal capacity $c_P$ and the following fitting law can be used for the thermal conductivity $\lambda$:

$$\lambda = 0.0818 + 2.76 \times 10^{-4} \times (T - 273.15) + 0.24 \times w \tag{5.4}$$

with $T$ the temperature in K and $w$ the water content in g $g^{-1}$.

### 5.3.1.2 Hygric Properties

Sorption isotherms are determined for samples of hemp, lime binder and sprayed HC by a static and discontinuous method according to ISO 12571 (2013). In order to get the adsorption (resp. desorption) curves, 8 $cm^3$ samples were placed consecutively in a series of test environments at constant temperature (≈23 ± 1 °C) with relative humidity controlled with saturated salt solutions increasing (resp. decreasing) in stage. Water contents are evaluated according to Eq. (5.3) once equilibrium is reached at each stage. Adsorption and desorption curves for sprayed HC are plotted on Fig. 5.6. In order to improve the legibility of the plot, experimental sorption data are described thanks to the GAB model (Künzel 1995) (see Fig. 5.6):

$$w = \frac{w_n C K \varphi}{(1 - K\varphi)\,(1 + K(C-1)\varphi)} \tag{5.5}$$

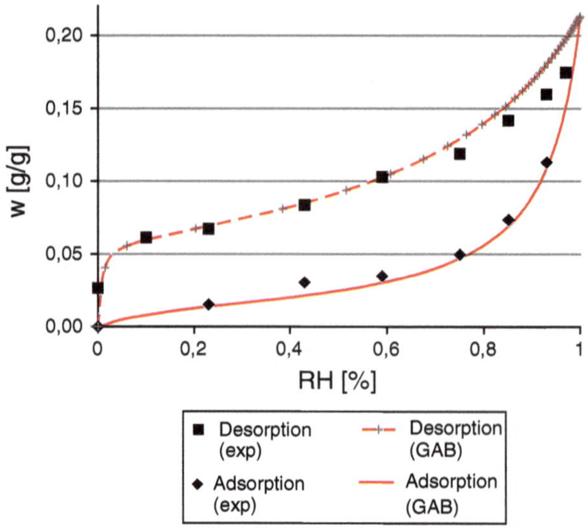


**Fig. 5.6** Adsorption and desorption curves of sprayed HC and fitting curves using GAB's model

**Table 5.3** GAB's parameters for adsorption and desorption curves

| | Adsorption | Desorption |
|---|---|---|
| $w_n$ | 0.015 | 0.06 |
| K | 0.93 | 0.72 |
| C | 12.5 | 200 |

where $w_n$, $C$ and $K$ are fitting parameters. Values for sprayed HC are summarized in Table 5.3. In addition, free saturation water content $w_{sat}$ was determined by immersing a sample in liquid water at room temperature. For a sample with a dry density of 425 kg m$^{-3}$, $w_{sat}$ is equal to 1.23 g g$^{-1}$.

The moisture transport mechanisms relevant to calculations in building physics are vapour diffusion and liquid transport through capillary forces (Künzel 1995), which are both characterized by a transport coefficient, namely the water vapour permeability and liquid water diffusion coefficient.

Water vapour permeability characterises the ability of a material to transfer moisture under a vapour pressure gradient. It is usually determined according to two standardised methods: EN ISO 12572 and ASTM Standard E 96/E 96M (2012) Mukhopadhyaya et al. (2007). The procedure is the following: a sample of known area $A$ and thickness $e$ separates two environments that differ in relative humidity under isothermal conditions and the rate of moisture flow $g_v$ (kg m$^{-2}$ s$^{-1}$) across the specimen under steady-state conditions is gravimetrically determined. Water vapour permeability $\delta$ is defined as:

$$\delta = \frac{g_v e}{\Delta p_v} \tag{5.6}$$

where $\Delta p_v$ is the water vapour pressure difference across specimen surfaces. The vapour permeability is usually replaced by the vapour resistance factor μ [−] as typical vapour transfer property of the building material:

$$\mu = \frac{\delta_a}{\delta} \tag{5.7}$$

with $\delta_a$ the air permeability, generally calculated by Schirmer relation as:

$$\delta_a = 2.306 \times 10^{-5} \frac{M_w p_0}{RTp} \left( \frac{\mathrm{T}}{273.15} \right)^{1.81} \tag{5.8}$$

where $M_w$ is the water molar mass, R is the universal gas constant and p and $p_0$ are the atmospheric and reference air pressure, respectively.

According to Mukhopadhyaya et al. (2007), various significant corrections (i.e. correction for resistance due to still air and specimen surface, edge mask correction, etc.) were taken into account in the experiment analysis to improve the results reliability. In this work, only dry cup experiments are performed to get the permeability at a mean relative humidity of 25 %rh. At this level, moisture is supposed to be transported only in the vapour form. Dry vapour resistance factor $\mu$ for sprayed HC is equal to 5.6 at 23 °C.

At higher moisture levels, liquid diffusivity can be evaluated through different complex and expansive methods (i.e. electrical methods, gamma-ray attenuation, neutron radiography, magnetic resonance, X-ray projection method, inverse method (Zaknoune et al. 2013), etc.). On the contrary, Krus and Künzel (1993) suggest evaluating liquid diffusivity simply through water absorption experiments as:

$$D_l^w = \left( \frac{A_w}{\rho_{dry} w_{sat}} \right)^2 \tag{5.9}$$

where $A_w$ is the water absorption coefficient in (kg $m^{-2}s^{-0.5}$), which can be determined from partial immersion experiments defined in the EN ISO 15148 (2002) as follow: a large surface is partially immersed in water during at least 24 h and the free water uptake is determined by measuring the mass increase. In the first stage of the experiment, a plot of the mass increase per surface unit against the square root of time is linear (Plagge et al. 2005). The slope of the line gives the water absorption coefficient $A_w$. For sprayed HC, $A_w = 2.2 \times 10^{-2}$ kg $m^{-2}$ $s^{-0.5}$ and $D_l^w = 1.9 \times 10^{-9}$ $m^2$ $s^{-1}$. This value is in the same order of magnitude than the one obtained from the Crank's analytical solution of Fick's diffusion law during early stage of drying (Colinart et al. 2012).

### *5.3.2 Constitutive Equations for Heat and Mass Transfer*

During the drying stage, water evolves through different states. At the beginning, free water forms a saturating or non-saturating continuous liquid phase (finucular state) and can be transported to the surface, mainly by capillarity, and evaporates until the pendular state is reached. At this point, liquid water exists in disconnected separate islands and water can only be removed via a slow vapour transport until the residual water content is reached.

The modelling of heat and moisture transport processes in hygroscopic porous material has been an ongoing concern in building physics and usually involves major theories of Philip and De Vries (1953), Luikov (1957) or Whitaker (1977). All are based on heat and moisture balance equations (i.e. Fourier's Law, Fick's Law, Darcy's Law) and describe moisture transport as an addition of vapour and liquid flow. Subsequently, several models for different applications have been developed considering the interdependence of heat and moisture within various wall components. The major difference between these models is related to the particular assumptions. In the case of drying of bio-based building materials, these include:

- Porous material is regarded as continuous, non-deformable (shrinkage is neglected) and stabilized (lime hydration and carbonation are neglected).
- Even if the material is heterogeneous at the microscopic scale, it is considered as homogeneous at the macroscopic scale.
- Local thermodynamic equilibrium is assumed at every point of the material.
- Gravity effects are neglected and total gas pressure is constant and equal to the atmospheric ones.
- Liquid water is supposed to wet the porous structure, i.e. $p_{cap} = p_g - p_l$.

Based on these assumptions, heat and moisture transport can be expressed in terms of temperature $T$ and mass water content $w$. Liquid convective flow is written using Darcy's, Kelvin's and Clapeyron's law in the capillary and hygroscopic domains as:

$$q_l = -\left(D_l^w \nabla w + D_l^T \nabla T\right) \tag{5.10}$$

$$D_l^T = D_l^w \frac{\varphi}{T} \frac{dw}{d\varphi} \ln \varphi \tag{5.11}$$

As there is no gas convective transfer, vapour transport occurs by diffusion. Fick's law, originally expressed as function of a concentration gradient, becomes:

$$q_v = -\left(D_v^w \nabla w + D_v^T \nabla T\right) \tag{5.12}$$

$$D_v^w = \frac{M_w p_v^{sat}}{RTdw/d\varphi} \frac{D_{v-air}}{\mu} \tag{5.13}$$

$$D_v^T = D_v^w \frac{\varphi}{T} \frac{dw}{d\varphi} \left(\frac{M_w L_v}{RT} - 1\right) \tag{5.14}$$

where $D_{v-air}$ and $p_v^{sat}$ are respectively the binary diffusion coefficient of vapour in air and the saturated vapour pressure.

Last, the water mass conservation equation can be described as:

$$\frac{\partial w}{\partial t} = \frac{\partial w}{\partial \varphi}\frac{\partial \varphi}{\partial t} = -\frac{1}{\rho_{dry}}\nabla.\ (q_v + q_l) \tag{5.15}$$

Due to the presence of low temperature levels, heat transfer has been attributed to conductive and convective effects only:

$$q_{th} = -\lambda \nabla T + q_l h_l + q_v h_v + q_a h_a \tag{5.16}$$

where $h$ is the enthalpy of liquid water, vapour or air.

For the energy balance equation, it comes:

$$\overline{\rho_{dry} c_P}\frac{\partial T}{\partial t} = -\nabla.(-\lambda \nabla T) - L_v \nabla q_v - (q_l c_{Pl} + q_v c_{Pv})\nabla T \tag{5.17}$$

where $\overline{\rho_{dry} c_P} = \rho_{dry}\left(c_{Pdry} + c_{Pl} w\right)$ is the heat capacity of the material.

To obtain the temperature and water content fields across the HC sample, the coupled and nonlinear partial differential equations were formulated in the "PDE Modes" of COMSOL Multiphysics® and solved based on an explicit scheme. Even if there are a large number of hygrothermal simulation tools (Delgado et al. 2010), COMSOL Multiphysics® offers flexibility in the equation and boundary conditions writing and may solve equations with variable time stepping (here the same as the experimental ones).

## *5.3.3 Results*

### 5.3.3.1 Simulation of Block Drying

Simulations are performed first for a sprayed HC block dried under forced convection with a mean air velocity measured in the air flue of 2 m s$^{-1}$. Boundary conditions are defined as follow: at the interface between the environment and the HC, we suppose the continuity of the heat flux and mass flux.

$$q_{th} = h\left(T_{ext} - T_{surf}\right) + L_v q_l \tag{5.18}$$

$$q_l + q_v = \frac{k_m M_w}{R}\left(\frac{p_v^{ext}}{T_{ext}} - \frac{p_v^{surf}}{T_{surf}}\right) \tag{5.19}$$

Convective heat transfer coefficient $h$ is set to 10 W m$^{-2}$ K$^{-1}$ for this experiment. Moisture transfer coefficient $k_m$ is calculated by Lewis's relation as:

$$k_m = \frac{1}{\rho_a c_{p,a} Le^{2/3}} \tag{5.20}$$

Under atmospheric circumstances, i.e., for most drying applications, $Le^{2/3} \approx 0.9$. Lateral sides are considered as adiabatic (for heat and mass transfer). No water is exchanged through the bottom sides, whereas temperature is set to the measured ones. Last, initial conditions are taken from the experiments.

Simulated mean water content, relative temperature at different location and humidity within the material are presented respectively on Fig. 5.7a–c. For the mean water content, a good agreement is found between experimental and numerical data over the entire drying stage. For readability, only the temperature at the centre of the material is plotted on Fig. 5.7b, even if measurements are available on other positions. Once again, numerical results follow accurately the observed experimental trend. Concerning the relative humidity, a difference is observed between experiments and simulation. Nevertheless, simulated relative humidity within the material starts to decrease at the same time that the experimental ones, indicating that the model is able to predict when the evaporation front reaches a certain region within the material. A further improvement could consist in evaluating the dependency of moisture transport coefficient to the water content or the relative humidity. In the case of building materials, Künzel (1995) takes into account the moisture dependency of suction liquid diffusivity with an exponential function:

$$D_l^w = 3.8\left(\frac{A_w}{w_{cap}}\right)^2 1{,}000^{\frac{w}{w_{cap}}-1} \tag{5.21}$$

For their part, Zaknoune et al. (2013) fit well their experimental results on bio-based render with the following function:

$$D_l^w = \exp\left(p_1 + \frac{p_2}{w}\right) \tag{5.22}$$

with $p_1$ and $p_2$ two parameters to be adjusted.

### 5.3.3.2 Simulation of Wall Drying

Simulations are now performed for a sprayed HC wall dried under natural convection within the bi-climatic chamber. Heat and mass transfer boundary conditions, defined in the previous section, are applied on interior and exterior side, whereas top and bottom are considered as adiabatic. For ambient temperature, convective heat transfer coefficient $h$ along a vertical surface is calculated to about $3\ \mathrm{W\ m^{-2}\ K^{-1}}$. Last, initial temperatures are taken from the experiments, whereas initial water content is uniformly set to $0.4\ \mathrm{g\ g^{-1}}$.

Simulation results are presented on (Fig. 5.8a, b) and compared to the temperature and relative humidity measured by a sensor located at $\approx 7 \pm 1$ cm of the interior surface.

Similarly to the block, temperature variations are well caught close to the interior chamber. For information, temperatures are also plotted at the wall centre (@ 18 cm) and close to the exterior chamber (@ 30 cm). Nevertheless, since

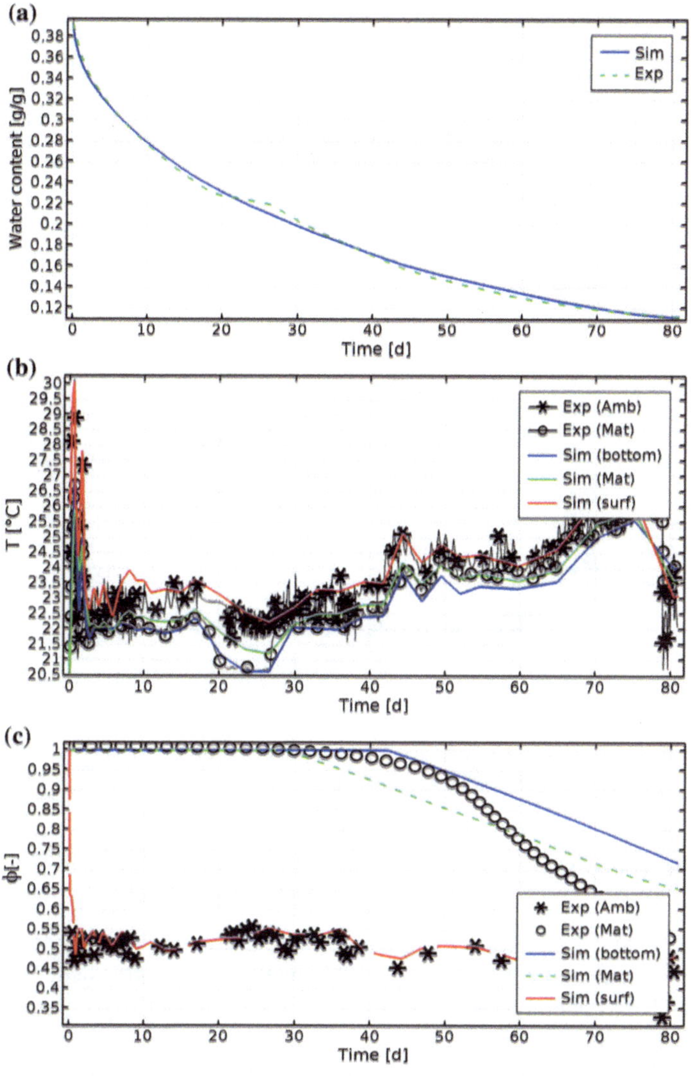


**Fig. 5.7** Comparison between experimental and numerical results of the drying stage of sprayed HC block: mean water content (**a**), temperature (**b**) and relative humidity (**c**)

temperatures are fluctuating in each chamber over the entire experiment, it is quiet difficult to observe isothermal stage linked to evaporation within the material. Therefore, relative humidity measurement is necessary. Once again, simulated relative humidity is in good agreement with the experimental one until the evaporation front reaches the sensor. Then (after the 60th day), a light deviation is observed. This point was previously discussed. Furthermore, relative humidity varies in a similar way because of symmetry of the set-up, whereas high relative humidity level is still observed in the wall centre, even after 100 days.

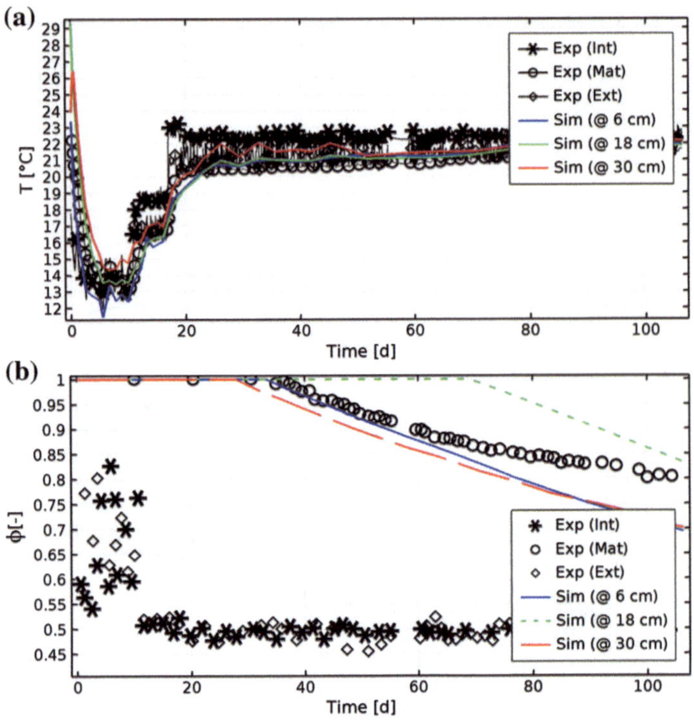


**Fig. 5.8** Comparison between experimental and numerical results of the drying stage of sprayed HC wall: temperature (**a**) and relative humidity (**b**)

## 5.4 Conclusions

Bio-based building materials like hemp concrete are still at a pioneer stage and still require many more efforts to gain knowledge on its transient hygrothermal behaviour. This chapter was focusing on the setting and drying stage of hemp concrete.

The results show first that the drying is an important stage of the manufacture of hemp concrete since it influence the setting of the porous, and thus of the mechanical, thermal and hydric properties. Then, it confirms the great influence of the setting process on the initial and final properties and on drying: spraying process leads to lower initial moisture content and lower density, and thus a faster drying time in comparison to the moulding or tamping process. Since this process is used in wet method of building, this result is advantageous in the view of applying render. On the other hand, it was shown that the initial hemp concrete formulation seems not to have an influence on the drying stage and thus on moisture diffusion. Last, it comes out that the block may be considered completely dried in terms of mass variation, whereas the relative humidity within the material is not equilibrated with its environment.

Furthermore, a simplified heat and moisture model was proposed by paying attention in the evaluation of the material properties. A roughly good agreement between experimental data and numerical results was found for the drying kinetic, temperature and relative humidity variations of hemp concrete blocks and wall.

The continuation of this work consists in refining the experimental and numerical analysis on the hygrothermal behaviour of hemp concrete, and particularly in the definition of the moisture transport coefficient.

**Acknowledgments** The authors want to thank the Brittany Regional Council, the National Research Agency of France and FEDER funds for their financial contributions.

## References

Green Building Home Page (2013) Available online: http://www.ciwmb.ca.gov/GreenBuilding/. Accessed 22 Apr 2013

Joseph P, Tretsiakova-McNally S (2010) Sustainable non-metallic building materials. Sustainability 2(2):400–427

Woolley T (2006) Natural building: a guide to materials and techniques, Ramsbury, The Crowood Press Ltd, Marlborough

Glass J, Dainty ARJ, Gibb AGF (2008) New build: materials, techniques, skills and innovation. Energy Policy 36(12):4534–4538

Karade SR (2010) Cement-bonded composites from lignocellulosic wastes. Constr Build Mater 24(8):1323–1330

De Bruijn PB, Jeppsson KH, Sandin K, Nilsson C (2009) Mechanical properties of lime-hemp concrete containing shivs and fibres. Biosyst Eng 103(4):474–479

Bütschi PY, Deschenaux C, Miao B, Srivastava NK (2004) Caractérisation d'une maçonnerie composée d'éléments en aggloméré de chanvre. Revue canadienne de génie civil 31(3):526–529 (in French)

Nguyen TT, Picandet V, Amziane S, Baley C (2009) Influence of compactness and hemp hurd characteristics on the mechanical properties of lime and hemp concrete. Eur J Environ Civil Eng 13(9):1039–1050

Nozahic V, Amziane S, Torrent G, Saïdi K, De Baynast H (2012) Design of green concrete made of plant-derived aggregates and a pumice–lime binder. Cement Concr Compos 34:231–241

Lanos C, Collet F, Lenain G, Hustache Y (2013) Formulation and implementation. In: Amziane S, Arnaud L, Challamel N (eds) Bio-aggregate-based building materials. Wiley, Hoboken 117–152

Chamoin J, Collet F, Pretot S, Lanos C (2011) Réduction du pouvoir absorbant de chènevottes par traitement imperméabilisant. Matériaux Tech 99(6):633–641

Lanas J, Pérez Bernal JL, Bello MA, Alvarez Galindo JI (2004) Mechanical properties of natural hydraulic lime-based mortars. Cem Concr Res 34(12):2191–2201

Lawrence RM, Mays TJ, Rigby SP, Walker P, D'Ayala D (2007) Effects of carbonation on the pore structure of non-hydraulic lime mortars. Cem Concr Res 37(7):1059–1069

Gu X, Song X (2010) Competition between hydration and carbonation in hydraulic lime and lime-pozzolana mortars. Adv Mater Res 133–134:241–246

Faure P, Peter U, Leseur D, Coussot P (2012) Water transfers within hemp lime concrete followed by NMR. Cem Concr Res 42:1468–1474

Thiery M, Baroghel-Bouny V, Bourneton N, Villain G, Stefani C (2007) Modelling of drying of concrete—Analysis of the different moisture transport modes. Rev Euro Gén Civ 11(5):541–577

Evrard A (2008) Transient hygrothermal behavior of lime-hemp materials. PhD thesis, Université Catholique de Louvain

Collet F, Chamoin J, Pretot S, Lanos C (2013) Comparison of the hygric behaviour of three hemp concretes. Energy Build 62:294–303

De Bruijn PB (2012) Material properties and full-scale rain exposure of lime-hemp concrete walls. PhD thesis, Swedish University of Agricultural Sciences

Colinart T, Glouannec P, Chauvelon P (2012) Influence of the setting process and the formulation on the *drying* of hemp concrete. Constr Build Mater 30:372–380

European Standard EN EN 772-13 (2000) Methods of test for mansonry units, Part 13: determination of net and gross dry density of mansonry units

Colinart T, Glouannec P, Pierre T, Chauvelon P, Magueresse A (2013) Experimental study on the hygrothermal behavior of a coated sprayed hemp concrete wall. Buildings 3:79–99

Cerezo V (2005) Propriétés mécaniques, thermiques et acoustiques d'un matériau à base de particules végétales : approche expérimentale et modélisation théorique. PhD thesis, ENTPE (in French)

Arnaud L, Gourlay E (2012) Experimental study of parameters influencing mechanical properties of hemp concrete. Constr Build Mater 28(1):50–56

Glouannec P, Collet F, Lanos C, Mounanga P, Pierre T, Poullain P, Pretot S, Chamoin J, Zaknoune A (2011) Physical properties of Hempcrete. Matériaux Tech 99:657–665

Elfordy S, Lucas F, Tancret F, Scudeller Y, Goudet L (2008) Mechanical and thermal properties of lime and hemp concrete ("hempcrete") manufactured by a projection process. Constr Build Mater 22(10):2116–2123

European Standard EN 12667 (2001) Thermal performance of building materials and products—determination of thermal resistance by means of guarded hot plate and heat flow meter methods—products of high and medium thermal resistance

Pierre T, Colinart T, Glouannec P (2013) Measurement of thermal properties of biosourced building materials. Int J Thermophys 1–21

EN ISO 12571 (2013) Hygrothermal performance of building materials and products—Determination of hygroscopic sorption properties

Guggenheim EA (1966) Application of statistical mechanics. Clarendon Press, Oxford

Künzel HM (1995) Simultaneous heat and moisture transport in building component: one- and two-dimensional calculation using simple parameters. PhD Thesis, Fraunhofer-IBP, Stuttgart

EN ISO 12572, Hygrothermal performance of building materials and products—Determination of water vapour transmission properties

ASTM E 96/E 96M (2012) Standard test methods for water vapour transmission of materials. ASTM international, West Conshohocken, pp 19428–2959

Mukhopadhyaya P, Kumaran K, Lackey J, van Reenen D (2007) Water vapor transmission measurement and significance of corrections. J ASTM Int 4(8):1–12

Zaknoune A, Glouannec P, Salagnac P (2013) Identification of liquid and vapour transport parameters of an ecological building material in early ages. Transp Porous Media 98(3):589–613

Krus M, Kunzel HM (1993) Determination of Dw from A-value. IEA Annex XXIV Report T3-D-93/02

EN ISO 15148 (2002) Hygrothermal performance of building materials and products—Determination of water absorption coefficient by partial immersion

Plagge R, Scheffler G, Grunewald J (2005) Automatic measurement of water uptake coefficient of building materials. In: Proceedings of 7th conference of building physics in Northern Countries, pp 15–22

Philip JR, De Vries DA (1953) Moisture movement in porous materials under temperature gradients. Trans Am Geophys Union 38:222–232

Luikov AV (1957) System of differential equation of heat and mass transfer in capillary porous bodies. Int J Heat Mass Transf 10:1–14

Whitaker S (1977) Simulation heat mass and momentum transfer in porous media: a dry theory of drying. Adv Heat Transfer 13:119–203

Delgado JMPQ, Ramos NMM, Barreira E, De Freitas VP (2010) A critical review of hygrothermal models used in porous building materials. J Porous Media 13:221–234

# Chapter 6
# Modelling of Wetting and Drying Cycles in Building Structures

**Delia D'Agostino**

**Abstract** Many buildings are frequently subjected to deterioration caused by dampness, whose effects can be amplified by structure material properties, and microclimatic conditions. This contribution proposes a quantitative model to perform a physical analysis of rising damp. The main driving forces which control the process were identified and expressed in formulae that involve generally obtainable microclimatic data. Wetting and drying cycles were analyzed to understand the periodic behaviour due to absorption and evaporation of moisture into and out of the structure. The model was applied to a field study to derive physical quantities that are not usually calculated in monitoring plans, such as the amount of water stored, the drying flow rate and the time-scale for drying. The daily evaporation rate was finally computed over a five year period and the harmonic equation fitting the curve was determined by considering meteorological data. The model showed that the drying capacity of the micro-environment was a primary factor in seasonal variations of the total flow through the building which lead to long-term damage. The developed model allowed much information on moisture cycles and on the associated deterioration of the structure to be obtained.

## 6.1 Introduction

Water presence in a structure is one of the main causes of building deterioration, but moisture dynamics are difficult to analyse, both theoretically and practically. Consequently, dampness mostly remains an unsolved concern.

Controlling moisture behavior in building enclosures has been a topic of growing interest especially over the last 15 years (Hall and Hoff 2012). However, a

D. D'Agostino (✉)
Scuola Superiore Isufi (SSI), University of Salento, via per Monteroni, 73100 Lecce, Italy
e-mail: delia.dagostino@unisalento.it

J. M. P. Q. Delgado (ed.), *Drying and Wetting of Building Materials and Components*, Building Pathology and Rehabilitation 4, DOI: 10.1007/978-3-319-04531-3_6,

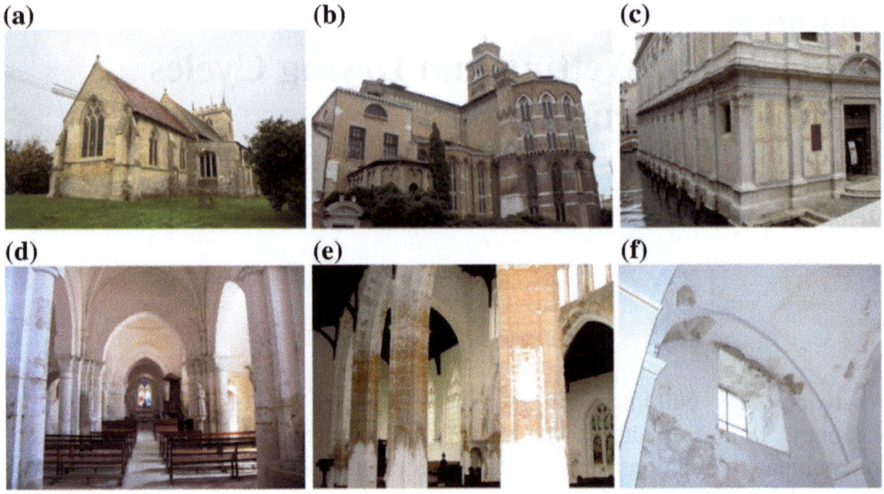

**Fig. 6.1** Examples of outdoor and indoor dampness in building structures: **a** Church of St. Mary in Chalgrove (Oxfordshire, England); **b** Basilica di Santa Maria Gloriosa dei Frari (Venice, Italy); **c** Church of San Giacomo Apostolo (Torricella, Italy); **d** Notre Dame (La Celle-Guenand, France); **e** St Andrew Church (Walpole, England); **f** Church of Santa Maria dei Miracoli (Venice, Italy)

comprehensive understanding of these dynamics is only rarely achieved due to the complexity of the physical and chemical processes involved (Moreno et al. 2006; Amoroso and Fassina 1983; Franzoni 2014).

The lack of this knowledge implies that in several buildings inadequate remedial measures have frequently accelerated damage (Freitas and Delgado 2013).

In the literature, as well as in practical observations, many cases are available to show how critical the consequences of dampness can became (Varas-Muriela et al. 2014). Moreover, stone deterioration due to moisture action can be strengthened in weathering of cultural heritage (Sawdy and Price 2005) (Fig. 6.1).

Wetting and drying cycles are an important deterioration factor in buildings and monuments (Franzen and Mirwald 2008; I'Anson and Hoff 1986). The periodical alternation between these cycles plays a key role in determining water movement and damage inside construction materials. Moisture regimes effects can be amplified in function of the specific features of a structure. For example, the drying of a wetted building element may rise when the indoor temperature increases and the relative humidity decreases at the evaporation zone (Rodriguez-Navarro and Doehne 1999).

Further concerns arise in heritage buildings as water is involved in the hydrolysis of silicate minerals, dissolution of carbonates and the formation of gypsum crusts (see Fig. 6.2).

An essential damage cause associated with dampness is linked to salt crystallization that can periodically occur in buildings in function of material properties and microclimatic conditions. Water can carry dissolved salts which cause decay by repeated crystallization, hydration and thermal expansion. Furthermore, an increase of salt efflorescence crystallization is favored when the evaporation rate

**Fig. 6.2** Rising damp damage in heritage buildings: **a** Völkerschlacht monument in Leipzig (Saxony, Germany); **b** Church of Sf. Arhangeli (Dolj Country, Romania); **c** Cave/Grotto of Paul and Thecla (Ephesus, Turkey); **d** tombs inside the Basilica Santa Maria Gloriosa dei Frari (Venice, Italy); **e** Kataluwa temple frescoes (Kataulwa, Sri Lanka); **f** Crypt of Lecce Cathedral (Lecce, Italy)

increases and water uptake decreases. In the opposite condition, efflorescence dissolution is favored when the evaporation rate decreases and water uptake increases (Arnold and Zehnder 1991).

A detailed physical analysis of moisture dynamics is, therefore, of crucial importance for buildings preservation (Guimarães et al. 2012).

The available computer-based tools aimed at understanding the masonry hygrothermal behavior often require a great deal of information, a high level of mathematical sophistication, and complex measurement instruments. Consequently, their application is difficult in practice and water transport modelling based on site surveys and measurements is not usually included in building monitoring and conservation activities (Straube and Burnett 1991).

Moisture dynamics can, however, be expressed by simpler models, able to describe the essential first order features of the process. These models avoid the use of computer based numerical methods but still emphasize the most important dynamics and their physical interrelationships.

## 6.2 Moisture Dynamics Theory

Modelling of water transport in building materials is a debated research topic. In the following paragraphs a schematization of rising damp in masonry is proposed, and then the sharp front (SF) theory is explained with the aim of modelling moisture dynamics.

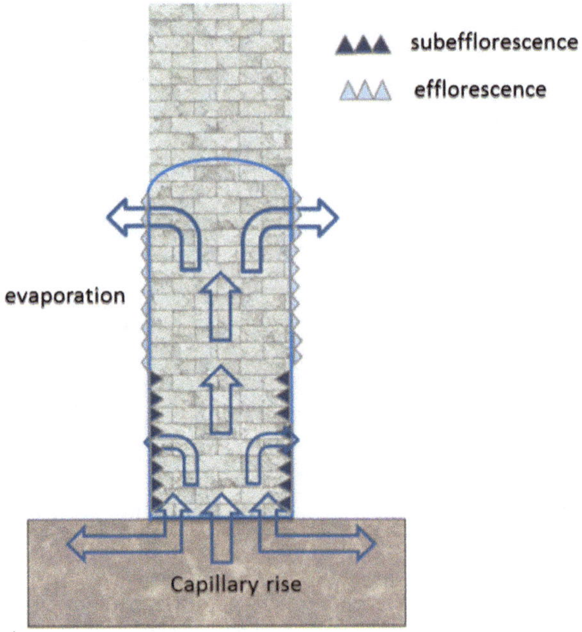


**Fig. 6.3** Rising damp in a masonry with the two main driving forces of the process: capillary rise and evaporation

### *6.2.1 Water Transport in Building Structures*

In the scientific literature, there are some studies that focus on the kinetics of capillary rise and other on rising damp features with a more qualitative and descriptive approach (Thomas et al. 1992; Massari and Massari 1998).

Many physical processes are involved in water transport in building structures. The main acting forces are hereafter synthesized.

Rising damp is the physical process through which moisture is transferred to and from masonry. In Fig. 6.3, the two main driving forces of the process, capillary rise and evaporation, are shown.

Water can reach the pore system of a structure by capillary suction of ground moisture, rain and condensation of air humidity. The process slowly carries moisture upwards from the lower parts of walls and other ground-supported structures leading to decay.

Capillary rise is particularly important when masonry is in direct contact with the ground as it can carry soluble salts (Torres and de Freitas 2007). Soluble ions, which are transported by rainwater or drawn in from adjoining materials, can enhance salt crystallization that gradually contributes to stone decay.

If crystallization occurs on the surface, efflorescence is formed where the evaporation rate of the environment is greater than the emission rate of the rising saline solution. On the other hand, subefflorescence may originate where crystals grow into the pore structure, and the emission rate of the rising saline solution is greater than the evaporation rate of the environment. This last case has an even more disruptive effect as it brakes internally the pore system of the material.

The process exacerbates when the water uptake carrying soluble salts is not balanced by the evaporation of the microenvironment.

Water is indeed lost from the structure turning into vapour by evaporation, which is influenced by many parameters, such as the moisture storage capacity of the structure and the microclimatic conditions at the site (Snethlage and Wendler 1996).

### *6.2.2 Unsaturated Flow: The Sharp Front Theory*

The sharp front (SF) moisture dynamics theory is reported to describe an analytical model of capillary rise in porous materials, incorporating both gravitational drainage and evaporation. The model calculates water movement in monoliths (single blocks of stone such as standing stone, gravestones and sculpture) and masonry structures, establishing several important scaling relations. The model is also helpful in case of geometrical complications and composite structures with different materials.

The physical investigation on rising damp is part the scientific theory of *unsaturated flows* (UF). The term unsaturated indicates that the materials within the building fabric are rarely fully saturated. If they were saturated, capillary forces would have been absent and water movement could only have happened in response to external forces, such as hydrostatic heads and gravity.

The mathematical models of these flows are quite complicated as they frequently involve hard numerical codes, computer based methods, and computations, with many variables difficult to be measured or estimated. Consequently, water transport site surveys, measurements as well as modelling are rarely performed during a building monitoring.

To overcome this issue, simplified models have been developed to express the essential first-order features of the processes. These models can avoid the use of hard numerical computations considering the important factors and their physical relationships with a reasonable estimation of the derived quantities.

An important class of simpler models is based on the sharp front (SF) theory that ignores the rather fuzzy boundary between wet and dry regions within a structure or fabric, and replaces it with a notional sharp boundary.

The model is referred to as lumped and it calculates how the boundary moves in the structure as water is fed into it from an external supply and is perhaps removed by evaporation. Hall and Hoff (2007) developed a physical model capturing in mathematical formulae the principal driving forces of rising damp.

In 2010 the authors modified it considering the evaporation process uniformly distributed throughout the wetted region (Hall et al. 2010). The authors showed how model predictions were consistent with empirical observations as the obtained formulae were accurate results of the simplified model, even if not representing exactly the phenomenon.

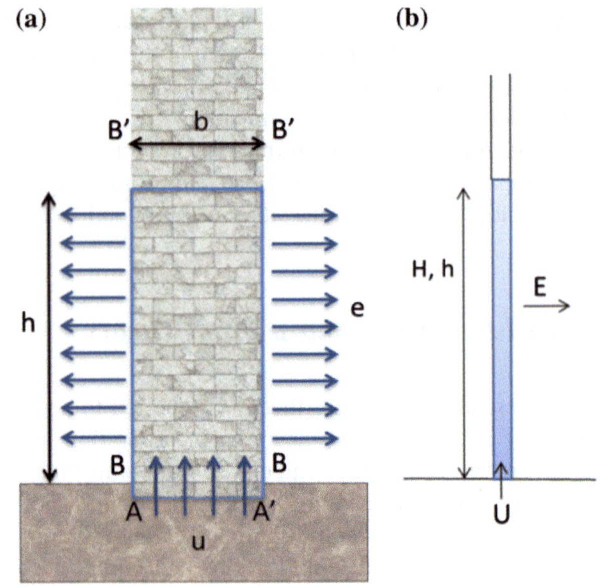


**Fig. 6.4** A schematized view of rising damp in a generic wall with the two main driving forces in the process. **a** A porous material of thickness $b$ is considered in hydraulic contact with a saturated ground at $AA'$. Water is absorbed at a rate $u$ per unit width of wall. The faces $BB'$ are subject to an evaporation rate $e$ per unit height of the wetted wall. The wetted height measured from the ground surface is $h$. **b** The lumped SF model: the total inflow $U = bu$ is controlled only by transport properties of the wall so that the total evaporation depends on the dimensionless height of the wetted region, $H$

### *6.2.3 Rising Damp*

The process of rising damp is now schematized with reference to a structure in hydraulic contact with the ground and of unlimited height.

In Fig. 6.4 the physical process of rising damp is shown in a structure composed of a porous material with a constant thickness $b$.

Rising damp is the result of the competition between the capillary absorption of water ($u$) through the boundary (AA′ in Fig. 6.4a) and the evaporation ($e$) of water along the exposed surface (BB′ in Fig. 6.4a) (Hall and Hoff 2007).

The total rate of absorption through the boundary AA′ is indicated by $U$ and the total evaporation rate by $E$.

In Fig. 6.4b, the lumped one-dimensional form of the model is shown. This model is useful to understand the scaling relations in the dynamics of rising damp. $E$ depends only on the height of the wetted region $h$, so whether evaporation occurs on one or two sides of the physical wall is not of interest.

Capillary forces $U$ work within the pores of the materials and are responsible for the initial uptake of water from external sources (Fig. 6.4a), such as groundwater, leaks or condensation, driving rain and splash water. They also cause migration of water within the fabric, and the redistribution of water from place to place usually associated with local differences in water content. In this process, capillary forces are generally dominant within the materials, so gravity can be omitted. However, Hall and Hoff also developed a mathematical model which includes the effects of gravity (Hall et al. 2010).

The water entering a structure through the foot of a wall depends on the capillary water absorption properties of the wall material(s). Only in saturated materials capillarity would be absent and water movement could occur in response to hydrostatic heads and gravity.

For almost every construction material, the cumulative volume of water $i$ absorbed (per unit area of inflow surface) into a dry material can be calculated as:

$$i = \theta_w h = St^{1/2} \tag{6.1}$$

where $\theta_w$ is the moisture content of the wetted region $h$ of the wall or, more precisely, the volume of water per unit volume of material, averaged over the entire wetted region, $t$ is the elapsed time and $S$ is the material sorptivity.

The sorptivity ($S$) is not an empirical parameter, but a property of a material rigorously defined in the theory of unsaturated flow and the capillary diffusion theory based on the Buckingham–Richards equations. $S$ can be easily measured by a laboratory test on a small sample of material.

Evaporation may occur from a building surface or inside the material followed by vapour migration before an eventual release into the atmosphere (Hall et al. 1984). The total rate of evaporation $E$ depends on the height of the wetted region $h$ and the potential evaporation $e$ established by the microenvironment, as:

$$E = eh \tag{6.2}$$

where $e$ is the evaporation rate of water per unit area of the free water surface (BB′ in Fig. 6.4a). The rate of evaporation of water from moist porous materials is determined by the environmental conditions over a wide range of water contents, from saturated through to rather dry. The quantity $e$ is rarely directly measured and it is influenced by temperature, relative humidity, air flow speed at and close to the wall surface.

The capillary absorption $U$ varies linearly from the inflow boundary at height $z = 0$ to the wet front at height $z = h$, so that:

$$U - U_O = -ez \tag{6.3}$$

with $0 \leq z \leq h$.

Thus, the total rate of flow into a wall of thickness $b$ through the boundary is equal to:

$$U_0 = \frac{bs^2}{2\theta_w h} + \frac{eh}{2} \tag{6.4}$$

This formula means that water is lost uniformly from the wetted region and the capillary pressure potential varies non-linearly from the inflow face to the wet front. The equation also shows that the rate $U$ at which water is absorbed at the base of the wall varies inversely with the height of rise $h$.

At this point, two main conditions can be distinguished: the steady state and the dynamic state.

#### 6.2.3.1 Steady State Condition

The steady state condition occurs when the water that comes into a structure by absorption is equal to the water that goes out of it by evaporation.

When rising damp has stabilized, there is a steady state (denoted with the subscript *ss*) characterized by the balance between the water absorbed and the water evaporated, so that $Uss = Ess$. It is important to stress that this state is not a state of static equilibrium, but a dynamic state of balance between the two components. At this state, considering (6.1–6.4), the following equation can be derived:

$$E_{ss} = eh_{ss} = \frac{bS^2}{2\theta_w h_{ss}} + \frac{eh}{2} \tag{6.5}$$

where $h_{ss}$ is the stabilized height of rising damp.

That means that water is lost uniformly from the wetted region and the capillary pressure potential varies non linearly from the inflow face to the wet front.

The quantity $h_{ss}$ is equal to:

$$h_{ss} = \left(\frac{bS^2}{e\theta_w}\right)^{1/2} \tag{6.6}$$

From this formula, the steady-state height of rise varies with the square root of the thickness of the wall $b$ and inversely with the square root of the potential evaporation rate $E$ in the local microenvironment.

The height of rise $h_{ss}$ (m) also varies in direct proportion to the sorptivity $S$ of the wall material. The functional dependence of $h_{ss}$ on $b$ and $S$ agrees with the earlier analysis of Vos (1971).

Furthermore, a steady flow $F_{ss}$ of water through the system that, can be derived as:

$$F_{ss} = E_{ss} = eh_{ss} \tag{6.7}$$

The total quantity of water stored within unit length of the wall $Q_{ss}$ is given by:

$$Q_{ss} = \theta_w bh_{ss} \tag{6.8}$$

Moreover, the mean residence time of water in the wall $t_{ss}$ can be calculated as:

$$t_{ss} = \frac{Q}{F} \tag{6.9}$$

representing the mean journey time for a water molecule to travel through the wall.

#### 6.2.3.2 Dynamic State Condition

The dynamic state condition occurs when the water that comes into a structure by absorption is not equal to the water that goes out of it by evaporation.

If the absorption uptake and the evaporation loss are not in a balanced condition, the difference between the two quantities $U$ and $E$ causes the total quantity of stored water to change, so that:

$$dQ/dt = U - E \tag{6.10}$$

Since the water stored in the structure is $Q = \theta_w bh$, the following differential equation can be obtained:

$$\frac{\mathrm{d}h}{\mathrm{d}t} = \frac{S^2}{2\theta_w^2} \cdot \frac{1}{h} - \frac{eh}{2b\theta_w} \tag{6.11}$$

Equation (6.11) determines how the height of rise varies with time when capillary rise has not stabilized.

This equation can also be written more simply as:

$$h\frac{\mathrm{d}h}{\mathrm{d}t} = -ah^2 + c \tag{6.12}$$

where the quantities $a$ and $c$ are equal to: $a = e/2\theta_w b$ and $c = S^2/2\theta_w^2$.

For the initial value $h = 0$ at $t = 0$, the equation for the entire process of capillary rise process $h(t)$ can be written as:

$$h^2 = \frac{c}{a}\left[1 - \exp\left(-2at\right)\right] \tag{6.13}$$

From an initial dry state, the system reaches $h = 0.95\ h_{ss}$ in a time $t_{95} = 3b\theta_w/2e$, which is the time-scale for reaching rising damp steady state.

The time-scale $t_{95}$ is inversely proportional to the evaporation rate, and in poorly ventilated situations may be much longer.

Over long periods, when $t \to \infty$, the height of rise tends to equal $h_{ss}$: $h \to h_{ss} - (c/a)^{1/2} = S(b/e\theta_w)^{1/2}$ as in Eq. (6.6).

As the drying capacity of the local microenvironment represents a critical gap in the model assessment, evaporation rate computation has to be carefully considered.

### 6.2.4 *Evaporation*

Formulae equivalent in accuracy to the Penman equation are hereafter derived. The main advantage of the proposed derived formulae is that they can be used with available weather records such as temperature $T$, relative humidity $RH$, air speed $u$ and solar radiation $R_S$, which is indirectly estimated from bright sunshine hours $n$.

In a moisture dynamics study, evaporation rates have to be known but, as already pointed out, these measurements are almost never done, especially at a micro-environmental scale, as part of building surveys.

On the other hand, a large number of physical quantities are needed to derive evaporation from meteorological/environmental data. Therefore, this fundamental physical quantity is not usually estimated.

Evaporation ($E_{PEN}$) (mm/d) is mainly influenced by many parameters in a complex interrelation given by the Penman formula (1948, 1956, 1963):

$$E_{PEN} = \frac{\Delta}{\Delta + \gamma} \cdot \frac{Rn}{\gamma} + \frac{\gamma}{\Delta + \gamma} \cdot \frac{6.43(f_u)D}{\lambda} \tag{6.14}$$

where $Rn$ is net radiation at the surface (MJ/m$^2$/d), $\Delta$ is the slope of the saturation vapour pressure curve (kPa/°C), $\gamma$ is psychrometric coefficient (kPa/°C), $\lambda$ is the latent heat of vaporization (MJ/kg), $f_u$ is the wind function given by $f_u = a + b_u u$, where $a_u$ and $b_u$ are wind function coefficients that in the original equation are equal to: $a_u = 1$, $b_u = 0.536$, $u$ is wind speed (m/s) at 2 m height, $D = (e_s - e_a)$ is vapour pressure deficit (kPa), $e_s$ is saturation vapour pressure (kPa) and $e_a$ is actual vapour pressure (kPa).

A plethora of parameters are required to use this formula. However, only some of them (e.g. $T$, $RH$, $Rs$, or, more frequently, $n$, and $u$) are usually recorded in common weather stations. Other parameters, such as the radiation and vapour deficit, can be estimated indirectly from the available weather variables measured. Many others (e.g. the latent heat of vaporization, the saturation vapour pressure, the actual vapour pressure, the psychrometric coefficient, the slope vapour pressure curve, the atmospheric pressure, the effective emissivity of the surface, the clear-sky solar radiation, the Stefane Boltzmann constant, the cloudiness factor) can be derived but this could lead to difficulty in the calculation partly because they could be expressed in a variety of units.

Therefore, as the Penman formula is difficult to apply and evaporation site measurements are rarely made in building surveys or conservation plans, various formulae have been derived and can be found in the scientific literature to more easily estimate evaporation rates (Linacre 1993; Allen et al. 1998).

Valiantzas (2006) proposed some formulae based on simplifications of the "standardized" Penman formula (6.14). Based on extensions made to the algebraic formulae, the author subsequently derived other simple algebraic expressions to estimate reference crop evapotranspiration (Valiantzas 2014).

The author accurately approximated the Penman equation dividing evaporation into two terms: the aerodynamic ($E_{aero}$) and radiation ($E_{radS}$, $E_{radL}$), that together compose the equation $E_{PEN} = E_{radS} - E_{radL} + E_{aero}$.

The following formulae were developed by the author together with others that can be found in the provided references:

$$\mathrm{E}_{rads} \approx 0.051(1 - \alpha)\mathrm{R_s}\sqrt{(\mathrm{T} + 9.5)} \tag{6.15}$$

$$E_{radl} \approx 0.188(T + 13).\left(\frac{R_s}{R_A} - 0.194\right) \cdot \left(1 - 0.00014(0.7T_{\max} + 0.3T_{\min} + 46)^2\sqrt{\frac{RH}{100}}\right) \tag{6.16}$$

$$E_{aero} \approx 0.049(T\mathrm{max} + 16.3)\left(1 - \frac{RH}{100}\right)(a_u + b_u u) \tag{6.17}$$

where $\alpha$ is the reflection coefficient or albedo, $Rs$ the measured or estimated incoming solar radiation (MJ/m$^2$/d), $R_A$ is the extra-terrestrial radiation (MJ/m$^2$/d), $T$ is the temperature (°C), $RH$ the relative humidity (%), $a_u$ and $b_u$ as previously.

The authors also propose a more simplified form for the Penman equation when wind speed data are missing. The formulae proposed are equivalent in accuracy to the Penman equation with the advantage that they can be used with routine climatic data, such as $T$, $RH$, $R_S$ and $u$.

The agreement between approximate and exact values for the formulae is reported in Valiantzas (2006).

All the needed terms to apply these formulae can be computed indirectly from available weather data following a standardized computational procedure proposed by Shuttleworth (1993) and Allen et al. (1998) and synthetically reported hereafter.

$Rs$ measurements may not be available but the parameter can be estimated using the following empirical relationship:

$$R_s = R_A\left(0.5 + 0.25\frac{n}{N}\right) \tag{6.18}$$

where $n$ is the measured bright sunshine hours per day (h), and $N$ the maximum possible duration of daylight (h). In this chapter, only the main simplified expressions for $R_A$, $N$ and $n$ are reported, but in Shuttleworth (1993) and Allen et al. (1994) a set of equations are derived when the latitude of the site and the number of the Julian day are given.

The additional parameter $R_A$ is given by:

$$R_A \approx 3N \sin(0.131N - 0.95\phi) \text{ for } \phi \succ \frac{23.5\pi}{180} \tag{6.19}$$

where $\Phi$ (radians) is the latitude of the site, positive for the northern hemisphere, varying from 0° to 53° and valid in temperate zones.

In tropical zones,

$$R_A \approx 118N^{02} \sin(0.131N - 0.2\phi) \text{ for } \phi \prec \frac{23.5\pi}{180} \tag{6.20}$$

The following set of equations can be used when the latitude of the site and the number of Julian day $J$:

$$N = \frac{24}{\pi}\omega_s \tag{6.21}$$

where $\omega_s$ is the sunset hour angle (radians) given by:

$$\omega_s = \arccos(-tg(\phi)tg(\delta)) \tag{6.22}$$

where $\delta$ is solar declination (radians) given by

$$\delta = 0.409 \sin\left(\frac{2\pi}{365}J - 1.39\right) \tag{6.23}$$

If there are no tables available, $N$ can also be derived from the following formulas:

$$N \approx 4\phi \sin(0.53i - 1.65) + 12 \tag{6.24}$$

where $i$ is the number of the Julian day or the rank of the month in monthly estimations. Finally $n$ can be approximated by:

$$n = \frac{1 - fsky}{N} \tag{6.25}$$

where $f_{sky}$ is the sky cover, varying from 0 for clear skies to 1 for cloudy skies.

These simplified formulae are almost of the same accuracy as the rather complex set of equations suggested by the standardized Penman model.

The equations can be used to estimate daily evaporation rate not only outdoors but also indoors to model the evaporation component of the rising damp process occurring at the dynamic condition.

In the next part of this chapter a field study is analysed to show how these formulae can be used in a real building.

## 6.3 Field Study

In this section the developed moisture dynamics theory is applied to a field study. This part is important to move from a theoretical approach to a practical experience and to show a possible approach to understanding water and moisture behavior in a real building. Physical quantities that are not usually calculated in monitoring plans are derived. Among them, there are the amount of water stored, the drying flow rate and the time-scale for drying.

### *6.3.1 The Cathedral and the Crypt of Lecce (South Italy)*

The application of the model has been carried out in the Cathedral and the Crypt of Lecce (South of Italy). Both churches are important cultural buildings placed in the historical center of Lecce and show crystallization of salt efflorescence on the walls and columns as the main dampness consequence.

The Cathedral (Fig. 6.5a) was built with its Crypt in 1114 by the Normans and it is located 3 m below the present ground level (D'Agostino 2011). The two buildings are directly connected through two stairways.

The Cathedral has a Latin crossing with nave and aisles divided by compound piers. The Crypt has a Greek cross plan symmetrical to a central axis with the nave

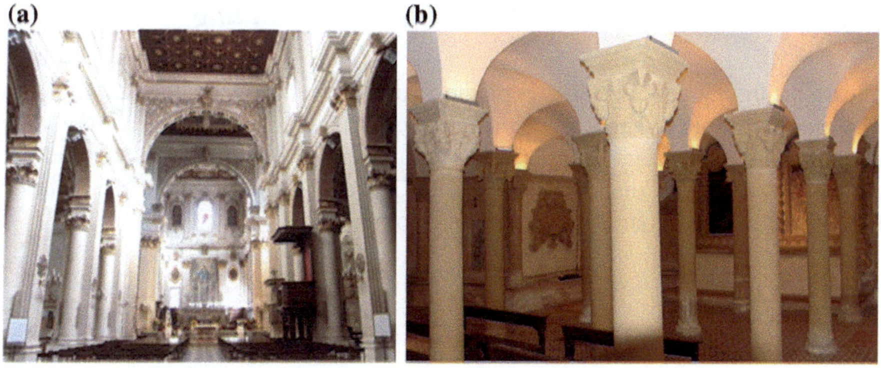

**Fig. 6.5** **a** The Cathedral and **b** the Crypt of Lecce (South Italy)

and the double aisles divided into equal parts by 42 monolithic columns (0.15 m radius), 4 double columns and 2 triple columns, which support a geometrical alignment of cross vaults (Fig. 6.5b).

The masonry walls are of close-jointed, fair-faced masonry of Lecce stone. Their thickness is approximately 1.5 m and the internal construction of the walls is of uncertain knowledge (Cataldo et al. 2012).

The masonry shows deterioration evidence in the form of white salt crystal efflorescence covering the walls, columns and the Baroque decorations. The Crypt displays problems as a direct or indirect result of moisture transfer through the masonry and the migration of soluble salts from the ground. In the building, the uptake occurs both from the subsoil and laterally through the perimeter masonry.

The Cathedral and the Crypt are made of Lecce stone, a typical soft limestone that is renowned for the property of absorbing water rapidly and drying slowly. It is characterized by a high porosity (30–40 %), capillary rise coefficient (9.4–11.8 $mg/cm^2\ s^{1/2}$), imbibition coefficient (13–13.9 %), a low dry index (0.040–0.067).

Therefore the construction material of the studied building is particularly susceptible to rising damp leading to the migration and accumulation of soluble salts that crystallize on its masonry.

Moreover, in the Crypt, capillary uptake is predominant as a water reservoir is maintained at the bottom of the walls and the columns (D'Agostino 2013).

Furthermore, the exterior surfaces of the perimeter walls are in contact with the soil that is a source of moisture from rainwater and groundwater, at least intermittently. This moisture contribute is remarkable above all during the rainy months. Another external water source can derive from infiltration, washing waters, pipes or drains.

The Crypt's position, the masonry layout (thick walls and many columns) and the physical properties of the stone favour capillary rise. But while the subsoil church shows relevant dampness, the above Cathedral benefits from the protective effect of the Crypt.

## 6.3.2 Moisture Dynamics Modelling in the Building

The moisture model formerly explained can allow the evaluation of the most important moisture dynamics quantities characterizing a building's masonry structure. The formulae previously derived are now applied both to the indoor and outdoor environments of the Cathedral and its Crypt. The methodology can be easily followed since the parameters required are commonly available data.

### 6.3.2.1 Steady State

In the first part of the research, the balanced capillary rise dynamic $Uss = Ess$ is considered in the application of the physical model to the building.

The Crypt masonry is hypnotized in a state of dynamic equilibrium between the total rate of absorption $U$ and the total evaporation rate $E$.

Equations (6.1–6.9) were used to derive the equilibrium moisture dynamics in the masonry of the building.

Before calculating $e$, some parameters have to be set in the model: the sorptivity and the moisture content of the wetted region. As regards $S$, in the literature of limestones, it usually lies in the range 0.5–1.5 mm/min$^{1/2}$. For Lecce stone the values vary between 0.7 and 0.9 mm/min$^{1/2}$ (Calia et al. 2002), or 0.4 and 0.5 mm/min$^{1/2}$ (Gombia et al. 2008) depending on stone variety. Capillary absorption experiments performed during this research on cubic samples of the same stone type of the Crypt of Lecce showed a range between 0.6 and 1.2 mm/min$^{1/2}$, so $S = 0.8$ mm/min$^{1/2}$ was used in the calculations.

The other important quantity to be defined for the model is the moisture content of the wetted region of the wall, $\theta w$. More accurately, it is the volume of water per unit volume of material, averaged over the entire wetted region. The quantity may be obtained by direct measurements of the moisture content in the wall. However, many previous studies have shown that $\theta w$ lies in a narrow range between the volume fraction porosity of the material $f$ (the conventional "porosity") and the so-called capillary moisture content of the material (the value obtained in short term capillary rise experiments).

Therefore, since the capillary moisture content is rarely less than approximately $0.7f$, taking $\theta w = 0.85f$ does not introduce much error. As the average volume fraction porosity $f$ of Lecce stone is between 30 and 40 %, $\theta w = 0.35$ was set.

A constant evaporation rate was calculated using the previous formulae and set equal to $e = 0.4$ mm/d per unit area of the wetted surface.

Equations (6.1–6.9) were modified to be applied to columns as water comes into the structure through a circumference. The total rate of flow into one column is obtained equal to:

$$U = \frac{\pi r^2 S^2}{2\theta_w h} + \frac{eh}{2}(2\pi r) \tag{6.26}$$

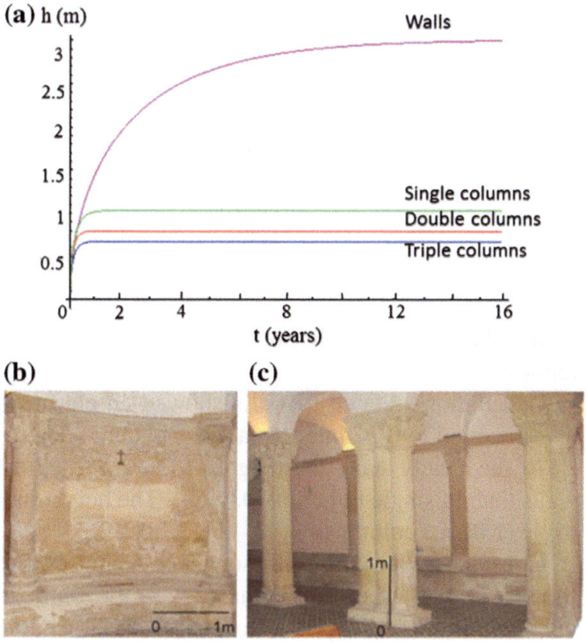


**Fig. 6.6** **a** The numerical solution of the height of rise $h$ as a function of time $h(t)$ in walls, single columns, double columns and triple columns. The evicence of the height of rise is enhanced by white salts efflorescence and it is clearly visible both on **b** the walls, and on **c** the double and triple columns of the Crypt

This gives a height of rise of:

$$h_{ss} = \left(\frac{rS^2}{2e\theta_w}\right)^{1/2} \tag{6.27}$$

Moreover, for the double and the triple columns of the Crypt, a 30 and 60 % circumference reduction was considered in the calculations of the evaporating surface (in Fig. 6.12b column's sections is showed).

More in detail, the double columns were considered as the result of the union of two columns with a 30 % reduced circumference, the triple columns as that of two columns with a 30 % reduced circumference (the laterals) plus one column with 60 % reduced circumference (the central).

The numerical solution of the height of rise as a function of time $h(t)$ (6.6) is shown in Fig. 6.6a in relation to the Crypt's walls and columns (single, double and triple).

In the walls, the height of rise $h$ as a function of time $h(t)$ is about 3 m, in single columns it reaches 0.7 m, while it goes up to 1 m in triple columns.

As shown by the curves slope in Fig. 6.6a, the system reaches the equilibrium more slowly in the walls than in the columns. This difference is explained by the different geometry of walls and columns. In the masonry structure, the increased thickness makes the system response slower.

In Fig. 6.6b, c the height of rise on the walls, double and triple columns is enhanced by the visible white salts efflorescence. The same figure displays how the height of rise in the building effectively reaches the predicted values.

**Table 6.1** Steady state dynamics in Crypt's masonry

| | Walls | Columns | | |
|---|---|---|---|---|
| | | Single | Double | Triple |
| $h_{ss}$ (m) | 3.02 | 0.68 | 0.81 | 1.07 |
| h (m) | 2.86 | 0.64 | 0.76 | 1.01 |
| $t_{95}$ (days) | 1,822 | 156 | 254 | 326 |
| $Q_{ss}$ (L) | 1,587 | 35 | 58 | 80 |
| $F_{ss}$ (L) | 1.3 | 0.3 | 0.7 | 1.2 |
| $t_{ss}$ (days) | 1,215 | 121 | 170 | 218 |

In Table 6.1 some other important parameters calculated for the steady state are reported.

Calculations related to walls showed an elevated quantity of water stored and an associated longer time to reach a moisture related steady state.

At the steady state, the height of rise $h_{ss}$ (3.02 m) is about four times more in walls than in single columns (0.68 m). The height increases in double (0.81 m) and triple (1.07 m) columns.

Considering an initial dry state, the height of rise $h = 2.86$ m is reached in a time $t_{95} = 1{,}822$ days in the masonry of the walls. In single columns, $h$ is 0.64 m while it increases to 0.76 m in double and 1.01 m in triple columns. The corresponding time $t_{95}$ is 156 days in single columns, increasing to 254 days and 326 days in double and triple columns.

The calculated values highlight how the moisture dynamics are particularly consistent in the building. In the walls, the water stored is $Q_{ss} = 1{,}587$ L with a water flow of $F_{ss} = 1.3$ L corresponding to a mean water residence time $t_{ss} = 1{,}215$ days. In single columns, the water stored is $Q_{ss} = 35$ L, increasing in double (58 L) and triple (80 L) columns.

The water flow $F_{ss}$ per day is 0.3 L in single columns, 0.7 L in double and 1.2 L in triple columns that respectively corresponds to a mean water residence time $t_{ss} = 121$ days for single columns, 170 days for double and 218 days for triple columns.

The moisture dynamics of the walls and of the columns present some important differences. In the case of the perimeter walls, whose exterior surfaces are in direct contact with the soil, the evaporation is not distributed along a surface parallel to the direction of flow, as it is in columns where water goes into the structure only through the base, but occurs on a surface at right angles to the flow. This is due to their particular construction largely below ground level with a lateral contribution of water absorption and infiltration.

For perimeter walls ($L = 1.5$ m), the maximum rate of capillary flow can be estimated using the sharp front analysis as $E_{max} = S^2/(\theta_w L)$. Finding $E_{max} = 1.7$ mm/d as the maximum rate possible at the interior wall face, these surfaces are likely to remain wet most of the year, drying back to produce efflorescence only when the $E$ is unusually high. However, there may be times during a typical year when the exterior soil moisture supply is restricted and then $E_{max}$ will fall correspondingly. Nevertheless, these calculations clearly show the general behaviour of the moisture regime in the structure of the building.

#### 6.3.2.2 Dynamic State and Evaporation Rate Modelling

Dynamic state modelling presents some common critical difficulties, such as quantifying evaporation directly or indirectly. In particular, the evaluation of the drying capacity of the microenvironment by evaporation measurements depends on many parameters difficult to be locally determined, while evaporation estimation using the Penman formula requires the knowledge of a large number of physical quantities. This section shows how these difficulties can be overcome using the proposed model.

In the dynamic state, the analytical solution of the differential equation of the moisture dynamics $\mathrm{d}Q/\mathrm{d}t = U - E$ (Eqs. 6.10–6.13) has to be found. As this can be difficult to obtain, a numerical solution was calculated using the software Mathematica.

The analytical solution of the differential equation for the height of rise was found by applying the lumped SF capillary rise model described previously. However, a feasible method to calculate how evaporation varies inside the Crypt and the Cathedral is required to solve the dynamic case.

Simplified formulae (6.15–6.17) were considered with the two evaporation components of the Penman equation.

The only contribution to the process in the indoors atmosphere was considered to be the aerodynamic term of the Penman formula: $E_{PEN} = E_{radS} - E_{radL} + E_{aero}$.

Equation (6.17) was chosen as a good approximation of the aereodynamic component of the Penman formula, with $a_u = 1$ and $b_u = 0.536$ (Valiantzas 2006).

The suggested empirical formulae (6.18–6.25) were used to calculate daily $R_S$, $R_A$, $N$, $n$, whilst $T$, $RH$ and $u$ measurements were automatically recorded every 360 min by an appropriate instrumentation placed outside and inside the Crypt and the Cathedral during the period from February 2011 to January 2012.

Platinum thermoresistance (Pt100, accuracy ±0.1 °C *T*) and thin-film capacitive sensors (accuracy ±2 % *RH*) were used to record *T* and *RH* while *u* measurements were taken using a hot-wire anemometer (accuracy ±0.1 m/s). Daily averaged data were considered in the calculations (D'Agostino 2010b).

*T*, *RH* inside and outside the Crypt and the Cathedral are reported in Fig. 6.7.

The graph shows how the outdoors influenced the indoors that, especially for *T*, suddenly followed the external variations. During the monitoring period the daily average outdoor conditions were between 7 and 33 °C for *T* and between 48 and 100 % for *RH*. The daily average *T* in the Cathedral was between 9 and 28 °C while in the Crypt it was between 10 and 26 °C. The Crypt's temperature was higher than the outdoors in February and lower than that of the Cathedral between March and November, with an average difference of 2.4 °C.

As regards *RH*, it was recorded between 52 and 88 % in the Cathedral while in the Crypt it was between 58 and 91 %. Aside from the *RH* decrease in the Cathedral between May and September, a net seasonal trend could not be found in the parameter that appeared very variable during the whole monitoring period.

As in the Crypt the indoor *RH* appeared to be quite high also in summer months when the outdoor water vapour contribution was rather low, other sources,

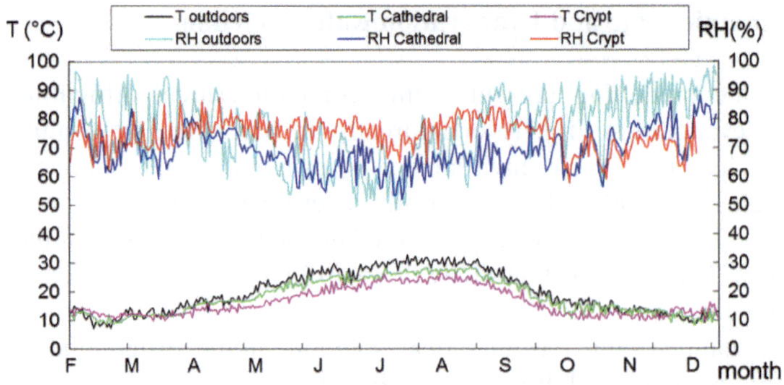


**Fig. 6.7** Average daily temperature (*T*) and relative humidity (*RH*) inside and outside the Cathedral and the Crypt in the period February 2011–January 2012

apart from the atmospheric input and mainly deriving from the soil, could be hypothesized.

*u* measurements gave average values of between 0.1 and 1.1 m/s.

Daily evaporation rates for the outdoor of the Cathedral and the Crypt of Lecce were calculated with the modified Penman equation using formulae (6.15–6.25).

Data (*T*, *RH*, *u*) recorded outside the Cathedral and the Crypt (latitude 40°21′06″N corresponding to $\Phi = 0.7$ rad) and tabulated data ($\alpha$, $n$, $N$) were used in the calculations (Atlante Climatico d'Italia 1971).

Results of the components $E_{aero}$, $E_{radS}$, $E_{radL}$ of $E_{PEN}$ are reported in Fig. 6.8a.

The graph shows how the main contribution to the outdoor evaporation is given by the heat term ($E_{radS}$, $E_{radL}$) while the aero tern ($E_{aero}$) has a minor influence.

The daily evaporation rate follows the heat term ($E_{radS}$, $E_{radL}$) trend that shows seasonal behaviour mainly influenced by temperature variations. The function fitting the *E* curve presents a typical hump-shape with maximal values from June to August (11 mm/d, 25/07/2011). Evaporation rate rapidly increases as temperature increases from March to July, while the opposite occurs from September to December. From December to January *E* outdoors reaches minima values (1.3 mm/d, 27/12/2011) when the difference between the heat and the aero terms is lower.

Figure 6.8b reports the graph showing the evaporation rate calculated indoors by the $E_{aero}$ component of the modified Penman equation using the microclimatic data (*T*, *RH*, *u*) recorded inside the Cathedral and the Crypt.

*E* values in the Crypt are always about half of those in the Cathedral.

The graph shows that both the Cathedral and the Crypt's indoors present a similar seasonal response with maxima in Spring and Summer and minima in Autumn and Winter of the monitoring period.

The function fitting the *E* curve has the same hump-shape in the Cathedral and the Crypt with a steeper shape in the uppermost church. Maxima *E* values are found between July and September (1.2 mm/d, 26/07/2011 Cathedral, 0.8 mm/d, 3/08/2011 Crypt) while from December to January *E* values are minima (0.4 mm/d, 12/12/2011 Cathedral, 0.2 mm/d Crypt, 19/12/2011).

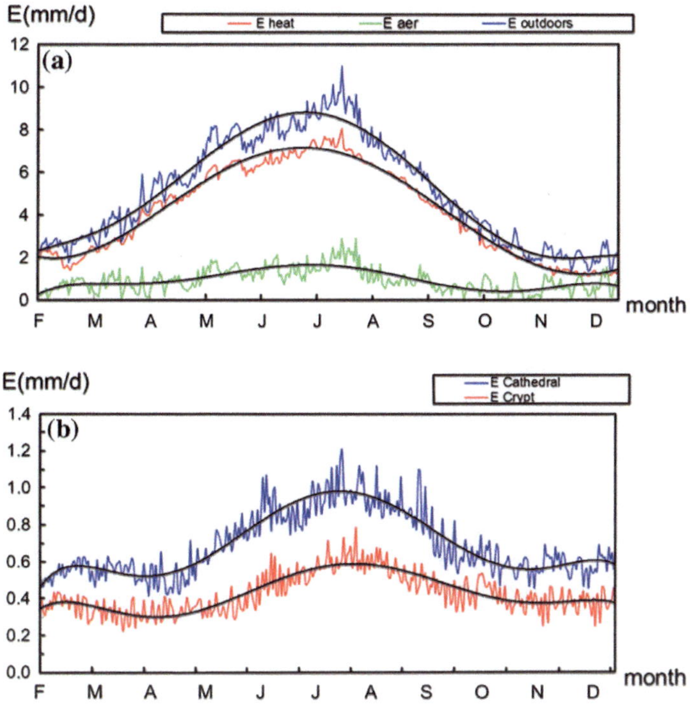


**Fig. 6.8** **a** Daily evaporation *E* (*blue curve*) outside the Cathedral calculated with the modified Penman equation using common meteorological data. The two main contribution terms, the *E heat* term (*red curve*) and the *E aero* term (*green curve*) are showed for 1 year. The polynomial functions (*black curves*) fitting the evaporation curves are also shown. **b** Daily evaporation *E* inside the Cathedral (*blue curve*) and the Crypt (*red line*) calculated for 1 year with the modified Penman equation (aero term) and common microclimatic data. The fitting curve (*black curve*) superimposed is also reported

### *6.3.3 Model Validation*

This section provides evidence of the model validation. A simple site method is now briefly described and compared with results from a modified Penman equation.

To test the accuracy of the derived formulae and validate the model, evaporation measurements were periodically made inside and outside the Cathedral and the Crypt, as shown in Fig. 6.9a that reports the site survey localization.

Lecce stone small blocks of $(0.2 \times 0.1 \times 0.05)\ m^3$ were used following an experimental procedure mentioned by Hall and Hoff but not yet used in buildings (Hall and Hoff 2012).

As example of a typical autumn day, results related to the measurements of 19/09/2011 (Crypt *T* 21.1 °C, *RH* 74 %, Cathedral *T* 23.6 °C, *RH* 66 %, outdoors *T* 25.4 °C, *RH* 77 %) are synthesized in Fig. 6.9b.

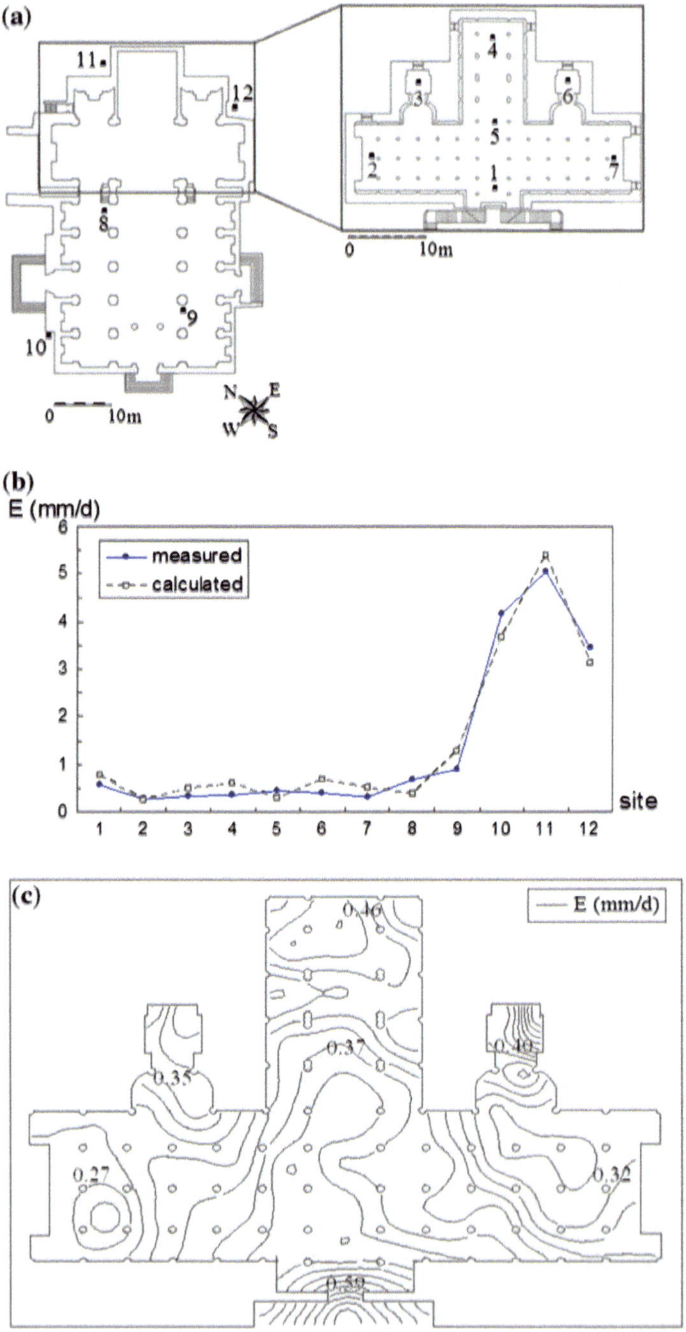


**Fig. 6.9** **a** Surveys localization inside and outside the Crypt and the Cathedral. **b** Measured and calculated evaporation (*E*) obtained by evaporation surveys and by the model. **c** Horizontal distribution of evaporation inside the Crypt obtained by interpolation (*Kriging* method) using ArcView Gis

Evaporation varies in the range from 0.3 and 0.6 mm/d in the Crypt.

In the Cathedral values were found between 0.7 and 0.9 mm/d while outside they were between 3.4 and 5.0 mm/d. The derived formulae gave 0.5 mm/d (Crypt's indoors) and 0.7 mm/d (Cathedral's indoors) and 4.7 mm/d (outdoors) as average results for the same day.

Measurements showed a good agreement with the values derived from the Eqs. (6.15–6.17).

Evaporation results were interpolated by the *kriging* method of ArcView Gis to estimate the $E$ distribution inside the Crypt.

In Fig. 6.9c the evaporation horizontal distribution is presented. The figure shows how $E$ is higher near the stairs that lead to the Cathedral where values reach 0.6 mm/d.

Values are also higher in the extremity of the main nave and near the masonry walls (0.5 mm/d). In the other parts of the building the distribution of the parameter is more uniform (interval between 0.2 and 0.4 mm/d) as seen by the evaporation isolines of Fig. 6.9c.

### 6.3.4 Estimating Potential Evaporation in the Microclimate

The daily evaporation rate is now computed over a five year period and the harmonic equation fitting the curve is determined by considering meteorological data.

By following the standardized sequence of computations for the Penman equation reported before (Eqs. 6.15–6.25), the daily evaporation rate was computed over a period of 5 years (from 2003 to 2009) using measured long term daily average climatic or meteorological data for the city of Lecce (Lecce Galatina Meteorological Station, 40°14′N 18°09′E).

The Fourier regression equations used in this study are reported in (6.26–6.29). They represent how the mean daily potential evaporation (*PET*) varies over the studied period.

This quantity describes the rate at which water evaporates from a free water surface. It can be measured directly by determining the rate of evaporation of water from open pans or calculated using meteorological data, using the Penman equation or its simplified versions.

The model can be written in the following multiplicative form:

$$PET_{p,t} = \overline{PET}_t \cdot E_{p,t} \tag{6.26}$$

where $PET_{p,t}$ is the analytically calculated daily *PET*, $\overline{PET}_t$ *is* the periodic mean of the daily *PET*, and $E_{p,t}$ is the remaining error, a dependent periodic stochastic component.

The periodic parts of Eq. (6.26) are approximated by a finite Fourier series,which may be formed as:

$$x_t = \bar{X} + \sum_{J=1}^{2} \left( A_j \cos \frac{2\pi jt}{\omega} + B_j \sin \frac{2\pi jt}{\omega} \right) \tag{6.27}$$

where $\bar{X}$ is the general mean of the $X_{p,t}$ series; $m$ is the number of significant harmonics, $\omega$ is the span of the periodicity ($\omega = 365$ for daily values), $A_j$, $B_j$ are the Fourier coefficients (J = 1, 2…m), and $t = 1, 2\ldots.\ \omega$. The authors showed that the number of the significant harmonics are only the first two, so $J = 2$.

After a series of mathematical operation that can be found in Kotsopulos and Svehlik (1989), the following equation can be obtained:

$$\overline{PET}_t = \exp\left(m_{pea} + \sum_{j=1}^{2}\left(A_j \cos\frac{2\pi jt}{\omega} + B_j \sin\frac{2\pi jt}{\omega}\right)\right). \tag{6.28}$$

where $m_{pea}, A_1, A_2, B_1, B_2$ are the coefficients to be determined and

$$E_{p,t} = \exp\left(\left(m_s + \sum_{j01}^{3}\left(A_j \cos\frac{2\pi jt}{\omega} + B_j \sin\frac{2\pi jt}{\omega}\right)\right) \cdot e^*_{p,t}\right) \tag{6.29}$$

where $m_s$ is a coefficient that can be estimated as constant, and $e^*_{p,t}$ is a term which includes a random error.

The obtained trend for Lecce is shown in Fig. 6.10a and the equation fitting the Lecce evaporation curve is reported in Fig. 6.10b. The model synthetically reported and discussed in Kotsopulos and Svehlik was followed while the curve was found by using the *FindFit* function of Mathematica.

In Fig. 6.10c the raw daily potential evaporation from Lecce is shown, with the regression curve superimposed and the parameters of the harmonic equation adjusted in Table 6.2.

Lecce's trend was compared with that of Athens and London (Temesgen et al. 1999; Yang et al. 2005). In both cases, the PE was calculated using the Penman equation.

Fourier regression equations were derived to represent the variation of the mean daily potential evaporation over a calendar year. The London data extend over a period of 5 years (1989–1994) and the Athens data over a 9 year period (1977–1985).

The time-varying evaporation rate (Fig. 6.10) calculated for the city of Lecce (mean 4.91 mm/d) from 2003 to 2009 showed a good agreement with the data obtained from a one year period investigation outside the Duomo (mean 4.63 mm/d) (Fig. 6.8).

Moreover, Lecce's values are comparable to those from the city of Athens (mean 4.77 mm/d).

As expected, the Lecce values are about 3.7 times greater those from London (mean 1.32 mm/d).

The comparison of the calculated evaporation raw data for Lecce with that derived by Castrignanò et al. (1997) for the South of Italy as well as with that obtained by site measurement also showed a good agreement.

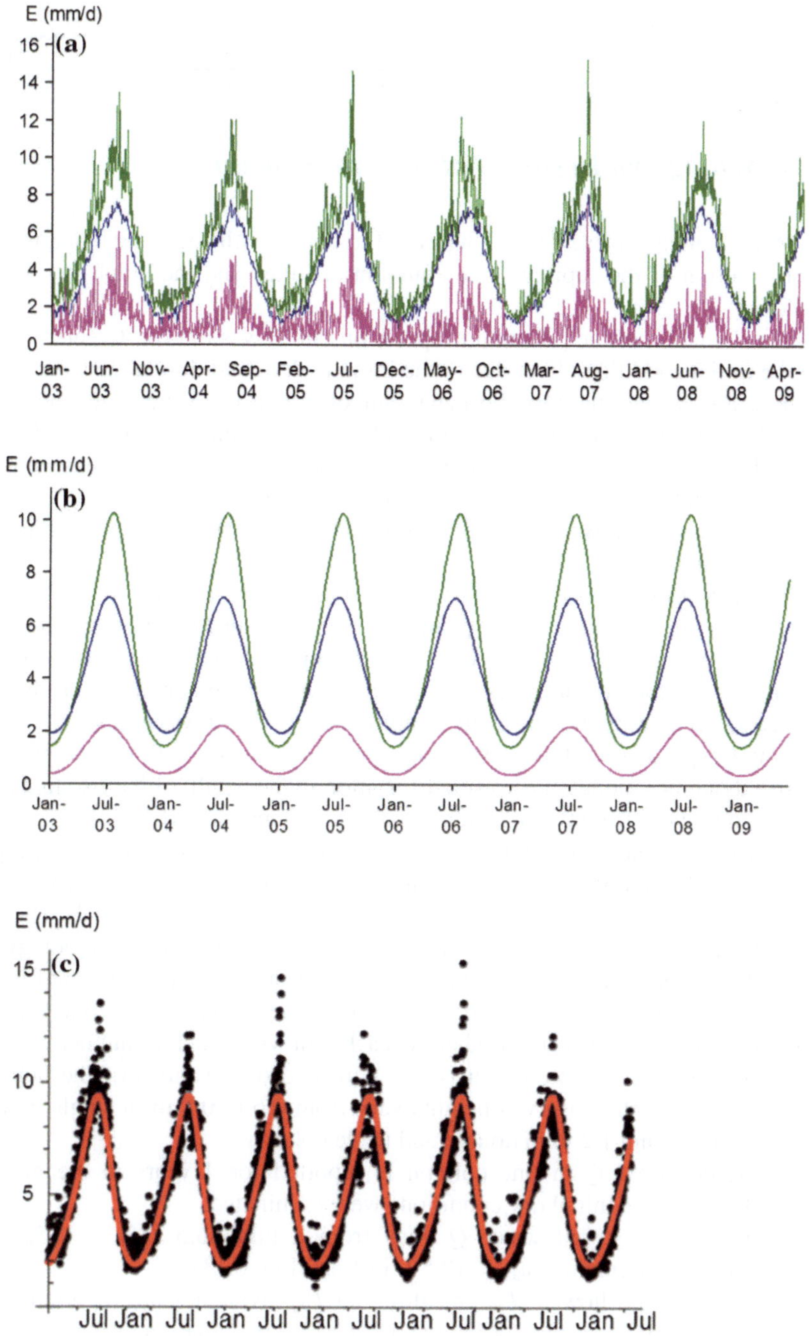


**Fig. 6.10** **a** Daily evaporation rate (*green line*) for the city of Lecce (years 2003–2009), showing the contribution of the aero term (*pink line*) and heat term (*blue line*). **b** The equations fitting the evaporation *Lecce curve*. **c** Raw daily potential evaporation from Lecce with the *regression curve* superimposed

**Table 6.2** The parameters of the harmonic equation calculated

| $A_1$ | $B_1$ | $A_2$ | $B_2$ | mpea |
|---|---|---|---|---|
| −0.7835 | 0.026478 | −0.022262 | 0.113491 | 1.45249 |

## *6.3.5 Wetting and Drying Cycles in the Structure*

The Mathematica solver *NDSolve* was used to compare the numerical and the analytical solution of the capillary rise equation obtained and the graph is shown in Fig. 6.11.

The calculation of the potential evaporation rate $E$ was run for 4 years to eliminate the initial condition effect due to the initially dry condition in the Crypt.

The graph shows the periodic behaviour of $E$ that every year increases from April (0.3 mm/d) to August (0.8 mm/d) and decreases from September (0.5 mm/d) to March (0.2 mm/d). The increasing of $E$ is more rapid than its decreasing with a maximum month variation from June to July ($\Delta e = 0.2$ mm/d).

In the columns, unlike the walls, evaporation is distributed along a surface parallel to the direction of flow.

The sharp front theory of the dynamic state can be applied to calculate the height of rise *h,* the water stored $Q$ and the total flow in *Fin* and out *Fout.* These quantities were derived using *NDSolve* eliminating an initial dry state of 4 years.

In Fig. 6.12a the height of rise $h$ is reported for a single column for a total time of 4 years, eliminating the initial dry conditions as in the previous case.

In the Crypt, with its lower evaporation rates calculated with the simple capillary rise dynamics model, the seasonal variation in the height of the rise in a single column goes about from 0.7 m in June to 0.6 m in September (Fig. 6.12a).

The height increases slowly from October to May while it decreases more rapidly from June to September (Fig. 6.12a). Moreover, as shown in Fig. 6.12b, the height of rise increases from singles (h1), to double (h2) and triple columns (h3). The same figure also highlights the seasonal variations in the different columns typology. The height of rise decreases in summer (blue line) and increases in winter (red line) in response to a higher or lower indoor evaporation. The crystallization zone is located between the levels that water periodically reaches in winter and in summer.

The model predictions are consistent with field observations of the height of rise (Fig. 6.6b) as the observed heights vary from 0.6 to 0.8 m in single columns and between 0.9 and 1.2 m in double and triple columns.

The water stored $Q$ in one column is reported for 4 years in the graph in Fig. 6.13a where the initial dry conditions were eliminated.

The volume of stored water $Q$ goes from a minimum $Q_{\min} = 15.5$ L in September to a maximum $Q_{\max} = 18.5$ L in June (Fig. 6.13).

In Fig. 6.13b the flow in $F_{in}$ and flow out $F_{out}$ dynamics are reported. These flows through one column underline the magnitude and variation of the total flow.

From July to September the flow out exceeds the rate of flow in while during the rest of the year the opposite occurs with the exclusion of a few months characterized by a rough balance between the two contributes (March, June and November).

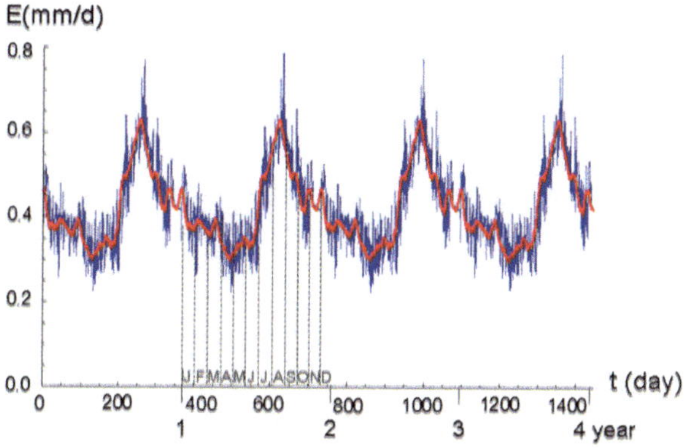


**Fig. 6.11** Numerical (*blue curve*) and analytical (*red curve*) daily potential evaporation calculated in the Crypt over 4 years to eliminate initial condition effects

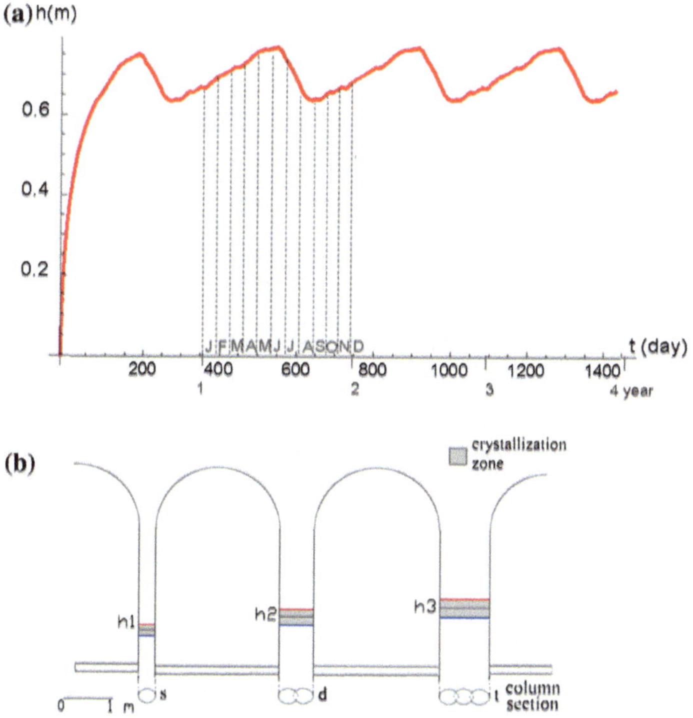


**Fig. 6.12** **a** Height of rise obtained using the Mathematica solver NDSolve for a column over 4 years to eliminate initial condition effects. **b** The average height of rise (*h1*, *h2*, *h3*) with seasonal variations (*red line* Winter, *blue line* Summer) in single (*s*), double (*d*) and triple (*t*) columns. The crystallization zone corresponds to the *grey area*

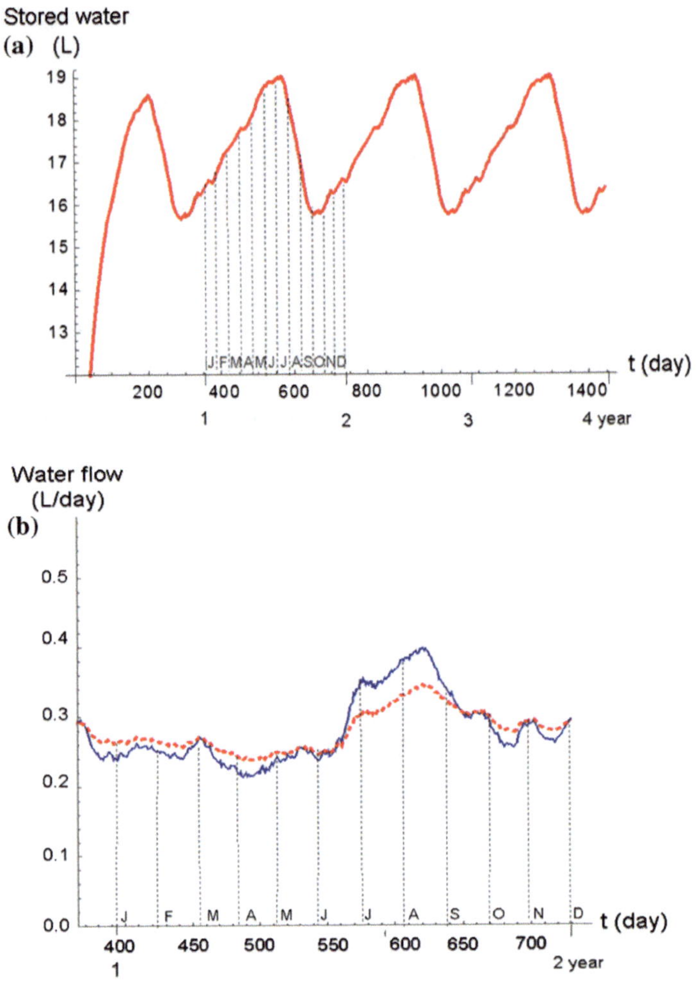


**Fig. 6.13** **a** Water stored obtained using the Mathematica solver NDSolve for a column over 4 years to eliminate initial condition effects. **b** Daily flow in (*dashed red curve*) and flow out (*blue curve*) in a typical year distribution for a column of the Crypt derived using the Mathematica solver NDSolve

This periodical behaviour is as a primary factor in long-term damage caused by salt crystallization and chemical degradation in the columns as the uptake of water is high and not balanced by the evaporation that appears close to free water.

To further verify the model predictions, the efflorescence diffusion on the Crypt's walls was periodically documented (D'Agostino 2010a).

During the period of microclimatic data acquisition, temporal and spatial variation of the salt-affected areas was estimated monthly using photographic documentation and site observations.

First, each photo was corrected for distortions using a specific tool, and then the graphic relief was converted into a digital format using a graphics software package (AutoCad). The surface covered by efflorescence was evaluated by image analysis tools developed in AutoCad and registered onto a photogrammetric base map. The comparison among the different maps of each month allowed the salts diffusion variations to be quantified.

This investigation confirmed that the relative abundance of efflorescence was apt to vary cyclically over the monitoring period.

The average area covered by efflorescence is reported in Fig. 6.14a with the recorded average airflow velocity (V). As the formula (6.17) indicates, this parameter is directly linked to the indoor evaporation rate that implies a major or minor water uptake by capillary rise. The dynamic equilibrium between the two processes has as a consequence an increase or a decrease of salt crystallization. Specifically, an increase of the evaporation rate implies an increased water uptake by capillary rise and an increase of salt crystallization (Arnold and Zehnder 1991).

Salts are mainly present in the form of nitre ($KNO_3$), probably attributable in origin to the organic matter on which the Crypt was built. In the ionic mixture, ions $K^+$ and $NO_3^-$ can be transported through the porous stone material by water from the ground into the stone. The single salt $KNO_3$ would always be precipitated according to the indoor *RH*, which never goes in the range 92–96 % that is the *RH* equilibrium for niter in the range 10–30 °C. A deeper explanation of the salt crystallization dynamics and relation to *T* and *RH* is given in (D'Agostino et al. 2013a).

Figure 6.14a highlights that the airflow velocity in the Crypt was between 0.1 and 1.1 m/s during the monitoring period, and that the surface covered by efflorescence increased when the airflow raised and vice versa.

The Crypt's masonries get a water uptake from the floor and the surrounding ground, promoted by the constituting stone properties.

The graphic documentation related to the efflorescence diffusion showed a variation of the salts efflorescence distribution on the building's walls, as reported in Fig. 6.14b. This figure displays how the relative abundance of efflorescence was greater in winter than in summer months.

The month characterized by the lowest efflorescence area was July (12 % of the total area) whilst the highest diffusion was found in December (about 21 % of the total area). The walls were not subjected to any form of mechanical cleaning during the period, and it was unlikely that the efflorescence had become detached and fallen to the floor. Thus, the most probable explanation for the observed variation is that the salts underwent phase transitions from crystalline to aqueous solution and vice versa, in response to a moisture transfer linked to microclimatic changes (Arnold and Zehnder 1991).

Between June/July and September, the flow out exceeds the rate of flow in, the efflorescence appeared to diminish as the surface temperature increased (about 24 °C with relative humidity 79 % in average), while between December and March (about 13 °C with relative humidity 72 % in average), when the surface temperature fell, salts crystallization was again more evident.

All these investigations confirmed how wetting and drying cycles are the main driving forces of deterioration in building structures.

(a)

(b)

(c)

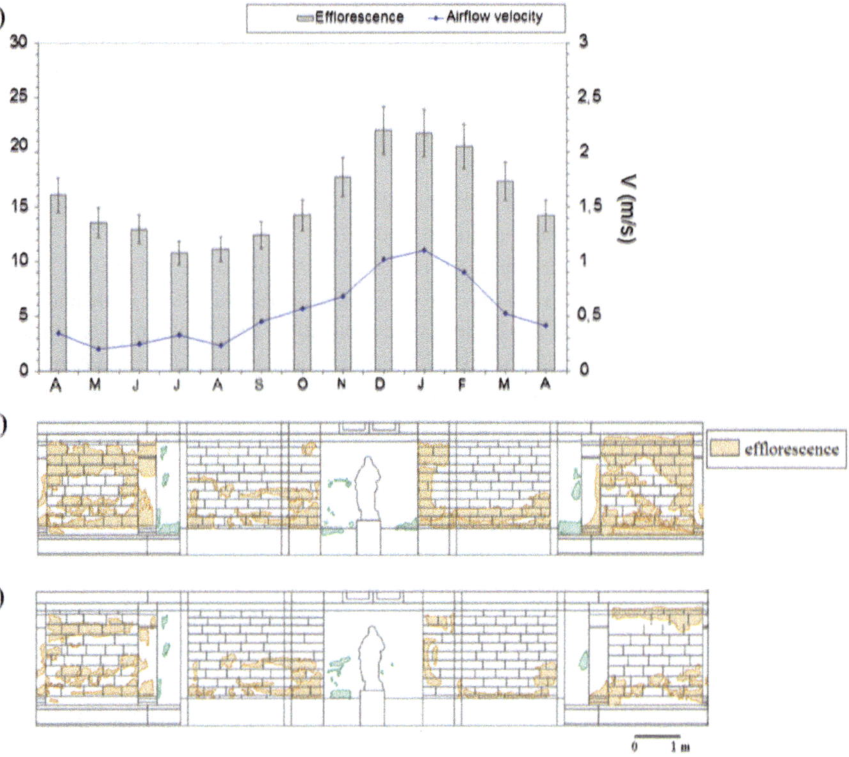


**Fig. 6.14** **a** Monthly area of the masonry covered by salts efflorescence with the standard deviation and the airflow velocity (*V*). **b** Winter and **c** Summer distribution of salts efflorescence

## *6.3.6 Moisture Cycles Evaluation*

The Crypt masonry gives rise to a unidirectional flow of moisture through the walls from the exterior to the interior, driven by the combination of capillarity and evaporation. As reported in (D'Agostino 2013), and further verified by computational fluid dynamic modelling of the indoor microclimate (D'Agostino et al. 2013b, D'Agostino and Congedo 2014), the long time-scale of water in the masonry and the low evaporation rate suggest that the Crypt might be poorly ventilated.

Furthermore, as calculated, the interior surface of the walls of the Crypt is subject to evaporation at a rate controlled by (and approximately equal to) the potential evaporation $E$ of the Crypt. As a consequence the wall surface remains wet. On the other hand, if $E$ exceeds this value, the capillary flow through the wall is unable to supply sufficient water to satisfy the potential evaporation of the microenvironment, and the consequently the surface dries out. Under these conditions salt deposition and efflorescence are periodically expected on the masonry. Moreover, Lecce's climate, with hot and dry summer and mild wet winters, encourage the dominance of salt weathering, localized especially at the top of the capillary fringe.

When the evaporation rates are increased in summer, the wetted area of the masonry is reduced but the water fluxes are increased. At low height of rise this causes a considerable increase in vulnerability to damage. The seasonal variation in evaporation rate is particularly significant for salt crystallization. The process takes place at the zone interested by seasonal variation of the height of rise (Fig. 6.12) while the lower parts of the structures are relatively undamaged.

Consequently, an increase of salt efflorescence abundance is favored from June to September. In this period evaporation rate increases, the water stored in the masonry decreases and the niter can crystallize on the surface. On the contrary, dissolution cycles of efflorescence are favored from October to May, when the evaporation rate starts to decrease, the water stored and the height of rise increases again.

Additionally, the application of a hard, stiff and high adhesion force plaster to some walls led to many problems favouring the accumulation of water in the stone and giving rise to a zone with increased dissolution and recrystallization of the alkaline salts within.

## 6.4 Conclusions

Deterioration caused by dampness is a frequent issue in building structures. In this chapter a quantitative model to perform a physical analysis of rising damp has been described.

The main physical quantities that characterize masonry (e.g. height of rise, timescale for drying, water stored and water flow) have been calculated in a representative building to show the feasibility of the model and of the derived formulae.

The model has also been tested to estimate the accuracy of indoor and outdoor evaporation rate calculations. Meteorological data were also used to compute daily evaporation rates over a five year period and to find the harmonic equation fitting the curve.

The model showed that the drying capacity of the micro-environment was a primary factor in seasonal variations of the total flow leading to long-term damage of the building. Additionally, the field study underlined how environmental conditions and water movement are strictly connected to evidence of seasonal deterioration, and how large volumes of water periodically move into and out of masonry contributing to water-driven damage processes.

As the proposed formulae involve routine climatic and microclimatic data, the developed model can be easy applied to obtain much information on moisture cycles and on the associated deterioration. It could be also utilised in monitoring plans both to forecast moisture cycles and help to plan building maintenance.

## References

Allen RG, Smith M, Pereira LS, Perrier A (1994) An update for the calculation of reference evapotranspiration. ICID Bull 43(2):35–92

Allen RG, Pereira LS, Raes D, Smith M (1998) Crop evapotranspiration: guidelines for computing crop water requirements. FAO Irrigation and Drainage paper, 56, Rome, pp 1–300

Amoroso G, Fassina V (1983) Stone decay and conservation—atmospheric pollution, cleaning, consolidation and protection. Elsevier, Amsterdam

Arnold A, Zehnder K (1991) Monitoring wall paintings affected by soluble salts. In: Cather S (ed) The conservation of wall paintings. The Getty Conservation Institute, Los Angeles, pp 103–136

Atlante Climatico d'Italia (1971–2000) vol 2, Servizio Meteorologico dell'Aeronautica Militare

Calia A, Laurenzi ML, Lettieri MT, Mecchi AM, Quarta G (2002) A methodology of sustainable monitoring of operations carried out on stone monuments. The case of Lecce Baroque churches. Research and Technology, pp 48–60, ARKOS

Castrignanò A, Di Bari V, Stelluti M (1997) Evapotranspiration predictions of CERES-Sorghum model in Southern Italy. Eur J Agron 6:265–274

Cataldo R, D'Agostino D, Leucci G (2012) Insights into the buried archaeological remains at the Duomo of Lecce using ground-penetrating radar surveys (Italy). Archaeol Prospection 19 (3). doi: 10.1002/arp.1423

D'Agostino D (2010a) A methodology of damage assessment in cultural heritage: Graphic decay mapping. In: Proceedings of ICAMS-3rd International Conference on Advanced Materials and Systems, Bucharest, pp 433–438. ISSN 2068-0783,

D'Agostino D (2010b) A methodology of damage assessment in cultural heritage: Microclimatic monitoring. In: Proceedings of ICAMS-3rd International Conference on Advanced Materials and Systems, Bucharest, pp 439-444. ISSN 2068-0783,

D'Agostino D (2011) Indagini fisiche per la diagnostica dello stato di conservazione della Cripta del Duomo di Lecce. PhD thesis, Edizioni Universitarie Romane, pp 1–240. ISBN 978-88-6022-173-5

D'Agostino D (2013) Moisture dynamics in an historical masonry structure: the Cathedral of Lecce (South Italy). Build Environ 63:122–133. http://dx.doi.org/10.1016/j.buildenv.2013.02.008

D'Agostino D, Macchia A, Cataldo R, Campanella L, Campbell A (2013a) Microclimate and salt crystallization in the Crypt of the Lecce's Duomo. Int J Architect Heritage: Conserv, Anal, Restor. doi:10.1080/15583058.2013.777862

D'Agostino D, Congedo PM, Cataldo R (2013 b) Computational fluid dynamics (CFD) modelling of microclimate for salts crystallization control and artworks conservation. J Cult Heritage. http://dx.doi.org/10.1016/j.culher.2013.10.002

D'Agostino D, Congedo PM (2014) CFD modeling and moisture dynamics implications of ventilation scenarios in historical buildings. Build Environ. doi: 10.1016/j.buildenv.2014.05.007

Franzen C, Mirwald PW (2008) Moisture sorption behaviour of salt mixtures in porous stone. Chem Erde 69:91–98

Franzoni E (2014) Rising damp removal from historical masonries: A still open challenge. Constr Build Mater 54:123–136

Freitas VP, Delgado JMPQ (2013) Hygrothermal behavior, building pathology and durability. Springer, Berlin, p 232. ISBN 978-3-642-31158-1. doi:10.1007/978-3-642-31158-1

Gombia M, Bortolotti V, Brown RJS, Camaiti M, Fantazzini P (2008) Models of water imbibition in untreated and treated porous media validated by quantitative magnetic resonance imaging. J Appl Phys 03:094913-1

Guimarães AS, Delgado JMPQ, de Freitas VP (2012) Rising damp in building walls: the wall base ventilation system. Heat Mass Transf 48(12):2079–2085. doi:10.1007/s00231-012-1053-3

Hall C, Hoff WD (2007) Rising damp: capillary rise dynamics in wall. Proc Roy Soc A 463:1871–1884

Hall C, Hoff WD (2012) Water transport in brick, stone and concrete. Taylor & Francis, London and New York

Hall C, Hoff WD, Nixon MR (1984) Water movement in porous building materials—VI. Evaporation and drying in brick and block materials. Build Environ 19:13–20. doi:10.1016/0360-1323(84)90009-X

Hall C, Hamilton A, Hoff WD, Viles HA, Eklund JA (2010) Moisture dynamics of walls: response to micro-environment and climate change. Proc Roy Soc 467(2125):194–211

I'Anson SJ, Hoff WD (1986) Water movement in porous building materials—VIII. Effects of evaporative drying on height of capillary rise equilibrium in walls. Build Environ 21:195–200. doi:10.1016/0360-1323(86)90030-2
Kotsopulos SI, Svehlik ZJ (1989) Analysis and synthesis of seasonality and variability of daily potential evapotranspiration. Water Resour Manage 3:259–269
Linacre ET (1993) Data sparse estimation of potential evaporation using a simplified Penman equation. Agric Forest Meteorol 64:225–237
Massari G, Massari I (1998) Damp buildings, old and new, ICCROM, Rome. English translation of Risanamento igienico dei locali umidi 1998; Hoepli Ed. ISBN 88-203-1231-X
Moreno F, Vilela SAG, Antunes ASG, Alves CAS (2006) Capillary-rising salt pollution and granitic stone erosive decay in the parish church of Torre de Moncorvo (NE Portugal)-implications for conservation strategy. J Cult Heritage 7(1):56–66
Penman HL (1948) Natural evaporation from open water, bare and grass. Proc Roy Soc London Ser A 193:120–145
Penman HL (1956) Evaporation: an introduction survey. Neth J Agric Sci 4:9–29
Penman HL (1963) Vegetation and hydrology. Technical communication no. 53. Commonwealth Bureau of Soils, Harpenden, England
Rodriguez-Navarro C, Doehne E (1999) Salt weathering: influence of evaporation rate, supersaturation and crystallization pattern. Earth Surf Process Landf 24:191–209
Sawdy A, Price C (2005) Salt damage at Cleeve Abbey, England. Part I: a comparison of theoretical predictions and practical observations. J Cult Heritage 6:125–135
Shuttleworth WJ (1993) Evaporation. In: Maidment DR (ed) Handbook of hydrology. McGraw-Hill, New York, pp 4.1–4.53 (Chap. 4)
Snethlage R, Wendler E (1996) Moisture cycles and stone degradation. In: Baer NS, Snethlage R (eds) Saving our cultural heritage. The conservation of historic stone structures. Wiley, Chichester, pp 7–24
Straube J, Burnett EFP (1991) Overview of hygrothermal (HAM) analysis methods. In: Trechsel HR (ed) ASTM manual 40-moisture analysis and condensation control in building envelopes, pp 81–89 (Chap. 5)
Temesgen B, Allen RG, Jensen DT (1999) Adjusting temperature parameters to reflect well-watered conditions. J Irrig Drain Eng 125(1):26–33
Thomas A, Williams G, Ashurst N (1992) Control of damp in old buildings. Technical publication no. 8. Society for the Protection of Ancient Buildings, London, UK
Torres MIM, de Freitas VP (2007) Treatment of rising damp in historical buildings: wall base ventilation. Build Environ 42:424–435
Valiantzas JD (2006) Simplified versions for the Penman evaporation equation using routine weather data. J Hydrol 331:690–792
Valiantzas JD (2014) Simplified forms for the standardized FAO-56 Penman–Monteith reference evapotranspiration using limited weather data. J Hydrol 505:13–23
Varas-Muriela MJ, Martínez-Garrido MI, Fort R (2014) Monitoring the thermal–hygrometric conditions induced by traditional heating systems in a historic Spanish church (12th–16th C). Energ Buildings 75:119–132
Vos BH (1971) Suction of groundwater. Stud Conserv 16:129–144. doi:10.2307/1505493
Yang C, Chandler RE, Isham VS, Annoni C, Wheater HS (2005) Simulation and downscaling models for potential evaporation. J Hydrol 302:239–254

# Chapter 7
# A Review of Drying Theory and Modelling Approaches

**Sandrine Geoffroy and Marc Prat**

**Abstract** This chapter presents an overview of the physics of slow quasi-isothermal drying in building materials and associated modelling techniques. The latter include the traditional continuum models as well as pore scale techniques such as direct simulations or pore network models. Overviews on some related problem such as the combined problem of evaporation, salt transport and crystallization are also given. The analysis of the current state of the art emphasizes the capillary liquid films in a real material as a major modelling issue, both within the framework of classical continuum models and pore scale models. Still more challenging is the development of drying models in the presence of salt crystallization, which represents a necessary step toward the modelling of damages induced by the salt crystallization in building materials.

**Keywords** Drying • Building materials • Modelling • Capillary liquid films

## 7.1 Introduction

Drying of building material has a long history since it has been found very long ago that drying could be used to convert clay into a useful building material. Drying processes are still nowadays an important step in the processing

S. Geoffroy
Laboratoire Matériaux et Durabilité des Constructions (LMDC), UPS, INSA, Université de Toulouse, 135, Avenue de Rangueil, 31077 Toulouse, Cedex 04, France

M. Prat (✉)
INPT, UPS, IMFT, Université de Toulouse, Avenue Camille Soula, 31400 Toulouse, France
e-mail: mprat@imft.fr

M. Prat
CNRS, IMFT, 31400 Toulouse, France

J. M. P. Q. Delgado (ed.), *Drying and Wetting of Building Materials and Components*, Building Pathology and Rehabilitation 4, DOI: 10.1007/978-3-319-04531-3_7,

of building materials. Related to drying evaporation from building material is also an important topic in relation with the control of moisture in building walls and rooms. Evaporation is actually considered as the only one important way in which water can leave a material. Main traditional building materials are wood, plaster, mortar, brick, stones and concrete. Less traditional materials, such as wood fiber mats, hemp, sheep's wool, and cellulose/recycled paper products, to name only a few are also increasingly used in modern constructions. Moisture transfers and evaporation in these materials have been less studied and represent interesting research subjects. Within the limited scope of this chapter, we will focus on generic aspects observed in brick and concrete typically with the idea that these generic aspects are also of interest for the study of less traditional materials. Wood is of course also an interesting building material. To a large extent, the material presented in this chapter also apply to wood but wood has also specificities owing to quite different microstructures and mechanical properties, compared to brick or concrete. The related specific aspects are discussed in several references, e.g. Perré et al. (2007) and references therein, and will not be covered in this chapter.

One can distinguish the fabrication steps from the situations with the material as an element of a completed building. The drying steps in the fabrication process generally involve the consideration of temperature gradients and possibly pressurization effects of the gas phase within the material. By contrast, evaporation from a building material in place in a building can be simpler to study, especially if one is mainly interested from the transfer between a wall and the inside of a building. In the simplest cases, the temperature variation across the material is weak and one therefore can focus on quasi-isothermal drying situations. For simplicity, we shall focus on quasi-isothermal evaporation processes in this chapter. In fact we will mostly consider the typical configuration encountered in many drying laboratory experiments. However, references addressing more involved drying situations will be indicated.

The focus is mainly on modelling. From an engineering standpoint, modelling of the drying process is generally developed within the traditional framework of the continuum approach to porous media, e.g. Bear and Bachmat (1990). The main problem with the continuum approach is that the physics is somewhat lumped into macroscopic parameters hiding the phenomena occurring at the pore scale. This is why alternate approaches can be useful. These notably include the direct simulations, e.g. Kohout et al. (2006), and the pore network models, e.g. Prat (2002, 2011). They can be used independently of the continuum approach to study specific points that cannot be addressed properly using the continuum approach or in conjunction with the continuum approach for computing the macroscopic parameters, which must be known for obtaining analytical or most often numerical solutions from the continuum models.

In addition to the control of moisture in building material and the relative humidity in adjacent rooms, evaporation processes are also key aspects regarding the damages induced by salt crystallization in buildings and monuments. These aspects will be briefly discussed in this chapter.

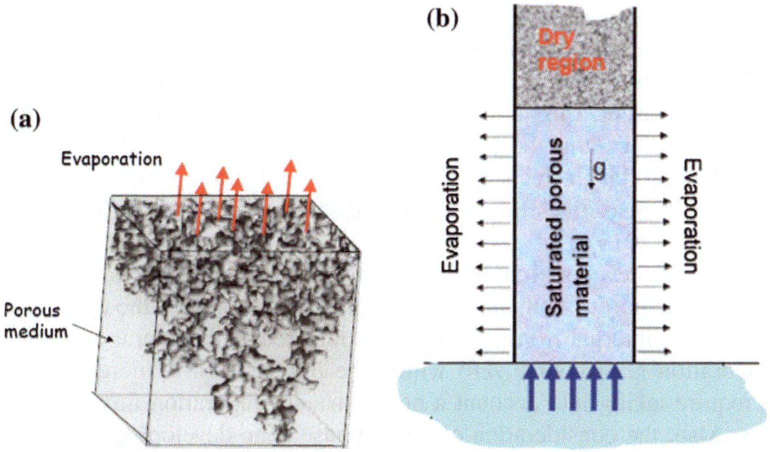


**Fig. 7.1** **a** Typical drying situation. **b** Schematic of the wicking-evaporation situation considered in Hall and Hoff (2002) and Veran-Tissoires et al. (2013)

The objectives of a drying or evaporation model are at least to predict the evolution of water mass loss and water content distribution within the material during the process. Depending on the situation under consideration, many other aspects can be of interest, such as the transport of species dissolved in the water as just evoked with the salt problem, or mechanical problems (cracks formation, deformations, shrinkage, etc.) or more generally quality problems (colour change, microorganisms development, etc.). Except the salt problem, all these interesting aspects will not be considered in this chapter.

Following Scherer (2004), two main evaporation situations can be distinguished in building physics, drying and evaporation—wicking. The two situations are schematically illustrated in Fig. 7.1. In drying, the limiting surfaces of the porous material are in contact with impervious walls or exposed to evaporation. In the evaporation–wicking situation, the sample is in contact at its bottom with an aqueous solution. The liquid is drawn into the pores by capillary suction. In contrast with drying, there is a permanent supply of solution and therefore a steady-state can possibly be reached when the flow rate of liquid sucked into the medium by capillary action exactly balances the evaporation rate. The result is the formation of an internal evaporation front as illustrated in Fig. 7.1. In laboratory experiments, the porous sample can be sufficiently short for the capillary action to maintain the medium fully saturated. Evaporation then takes place not inside but at the surface of the porous material. These three main different situations correspond to real situations, e.g. (Scherer 2004). For example, the fully saturated evaporation-wicking case can correspond to the section of a wall near the ground, whereas an internal evaporation front can exist at higher elevations along the wall. The drying case can apply to building materials located at still higher elevations, where the material is not connected to the ground water.

The situation of wick action with evaporation is considered in some details in Hall and Hoff (2002) and Veran-Tissoires et al. (2013) under the condition of a sharp front between the saturated region and the dry region. Thus, the corresponding material will not be repeated here. As reported in Veran-Tissoires et al. (2013), the resulting model, which is based on Darcy's equation, can be solved analytically. In brief, this model enables one to find the sharp front final position taking into account the evaporation process along the sides of the porous domain, the capillary effects and the viscous and gravity effects in the rising liquid phase. It also gives the dynamics of imbibition [see Eq. (8.24) in Veran-Tissoires et al. (2013)].

Note that the evaporation flux is assumed to be constant along the lateral boundary of the porous domain in the region below the front (the evaporation taking place at the internal front is neglected compared to the lateral evaporation and the evaporation is of course zero within the dry region). A more refined model would require taking into account a non-constant evaporation flux along the wall surfaces. Also, the consideration of a two-phase zone developing at the tip of the saturated region would be more realistic.

The remainder of the chapter focuses on the drying situation.

## 7.2 External Mass Transfer: Evaporation from Porous and/or Partially Wet Surfaces

Evaporation rates for both situations depicted in Fig. 7.1 are highly dependent on the external mass transfer, at least when the water content in the medium is sufficiently high. Thus a key issue is to model the external mass transfer (and of course also the heat transfer at the limiting surfaces of the porous medium when the assumption of quasi-isothermal evaporation cannot be made). The basic question is whether the evaporation rate from a porous surface, that is a surface which is partially solid and partially liquid, is different (and a *priori* smaller) from the same surface fully covered by liquid. This problem is often referred to as the Suzuki and Maeda problem after the first authors having addressed it, Suzuki and Maeda (1968). They essentially showed that the evaporation rate from a porous surface was identical to the rate from the same surface assumed fully covered by liquid provided the porous surface was sufficiently finely divided. In other terms, the pore sizes must be sufficiently small compared to the overall size of the surface for the rate being essentially identical to that of a liquid surface. This is of course when liquid occupies a sufficiently large number of pores at the surface. Note also that we implicitly supposed in this section that the pores are greater than 100 nm so that the Kelvin effect, see Sect. 7.5, can be neglected.

As an example Fig. 7.2 shows an image of the surface of a fired clay brick. Pores of few microns in size separated from each other by also about a few microns are clearly visible. This characteristic length a few microns denoted by $d$ is to be compared to a typical external transfer length scale $\delta$ (the thickness of the mass external boundary layer typically), which is typically on the order of 1 mm of greater. Thus $d/\delta$ is typically very small for building materials. In accordance

**Fig. 7.2** SEM image of a fired clay brick surface. Pores corresponds to the black spots

with Suzuki and Maeda findings this explains why it is acceptable to consider that the porous surface of a building material with liquid saturated pores leads to the same evaporation rate as the same surface entirely covered by a thin layer of liquid. The readers interested in more detailed considerations on the Suzuki and Maeda problem can refer to Shahraeeni et al. (2012).

The conclusion from this section is therefore that the boundary condition to impose at the surface of a saturated porous medium in order to compute the vapour concentration field in the external gas is the same as for a liquid surface, namely

$$p_v = p_{ve} \tag{7.1}$$

where $p_{ve}$ is the equilibrium vapour partial pressure at a liquid/gas interface. The evaporation flux is then given by

$$j = -\frac{M_v D}{RT} \frac{p_{atm}}{p_{atm} - p_{ve}} \nabla p_v \cdot \mathbf{n} \tag{7.2}$$

where $\mathbf{n}$ is the outwardly directed unit normal vector at the surface; $M_v$, $R$ and $T$ are the water molecular weight, the universal gas constant and the temperature; $p_{atm}$ is the total pressure in the gas phase and $D$ is the molecular diffusion coefficient of vapour in the binary mixture formed by air and the vapour.

To fully determine the external mass transfer, i.e. the field $p_v$ in the vicinity of the porous medium surfaces in contact with the external air it is in principle necessary to solve the external mass transfer problem, i.e. the P.D.E. governing $p_v$ in conjunction with the Navier-Stokes equation governing the velocity field in the external air. This can be of course not a trivial task for a complex geometry, especially if the external flow is turbulent. Going further into this direction leads to the consideration of drying as a fully coupled problem between the transfers in the external air and the transfers in the porous materials. This fully coupled approach to drying simulations, as well as some simplified approaches, is discussed for example in Defraeye et al. (2012).

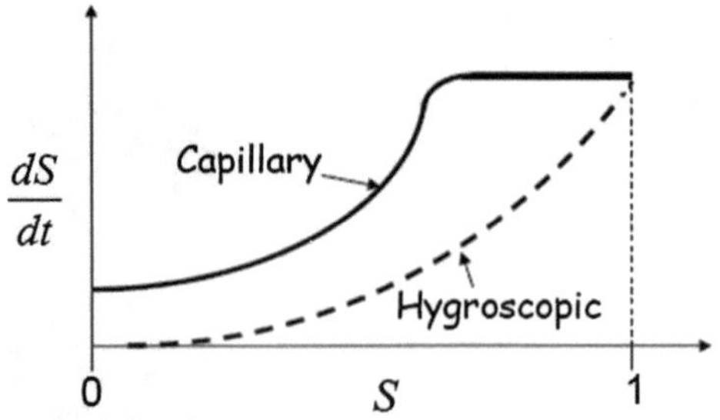


**Fig. 7.3** Schematic of typical (*slow*) drying *curves* for a capillary porous medium and a hygroscopic one. $S$ is the overall saturation and $t$ the time

In this chapter we will proceed simpler by assuming a classical boundary layer like expression of the form,

$$j = D\frac{M_v}{RT}\frac{(p_{vi} - p_{v\infty})}{\delta} \tag{7.3}$$

where $p_{vi}$ is the vapour partial pressure at the porous surface and $p_{v\infty}$ is the vapour partial pressure in the external air far away from the porous medium surface.

## 7.3 Capillary Porous Media Versus Hygroscopic Porous Media

Capillary porous media refers to porous materials in which the water is essentially fixed by capillarity. The fraction of the pore space that can be occupied by adsorbed water in a capillary porous medium is negligible. In terms of pore size distribution, this means that the fraction of pores that are less than about 100 nm is absent or negligible. Bricks can be considered as reasonably good examples of capillary materials. By contrast, the pore sizes will be all less than about 100 nm in purely hygroscopic materials. Cement paste can be considered as an example of hygroscopic material, i.e. a material in which a significant amount of water can be fixed by adsorption. Naturally the pore size can span over both the submicron range and the submillimeter range in some materials, leading to the consideration of partially hygroscopic capillary porous media. Concrete is an example of material with pores typically in both the capillary range and the hygroscopic range.

The distinction between capillary porous media and hygroscopic porous media is important in drying because they lead to quite different drying curves as schematically illustrated in Fig. 7.3. One obvious difference is the constant evaporation rate observed over a significant period of drying with a capillary porous medium, which is not obtained with a hygroscopic material. This is notably due to the fact that the vapour partial pressure at the surface menisci of a capillary porous media is not different from the equilibrium partial pressure at a flat liquid surface at the same temperature. By contrast, the vapour partial pressure is less at the surface menisci of a hygroscopic material because of the impact of menisci curvature on the equilibrium partial vapour pressure (see Sect. 7.5).

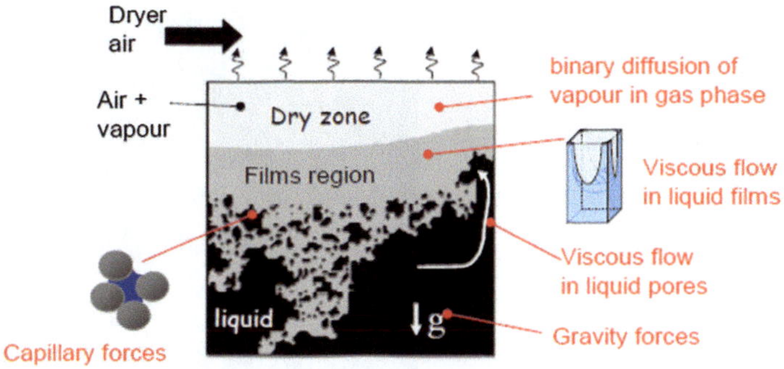


**Fig. 7.4** Main mechanisms controlling the isothermal drying of a capillary porous medium

## 7.4 Physics of Drying of Capillary Porous Media

### *7.4.1 Main Mechanisms*

The main mechanisms involved in the isothermal drying of a rigid non-deformable capillary porous medium are summarized in Fig. 7.4. The vapour transport is by diffusion in the gas phase considered as binary mixture of air and water vapour. The total gas pressure is assumed spatially uniform equal to atmospheric pressure at any time. The forces that can play a role in the evolution of the liquid distribution within the material during drying are the capillary forces, the viscous forces and the gravity forces. As discussed in Sect. 7.4.3, the evolution of drying rate can be highly dependent of the liquid films. The external mass transport, generally turbulent in the applications, controls the external evaporative demand.

### *7.4.2 The Main Drying Regimes*

Consider the typical drying situation depicted in Fig. 7.1a assuming a fully liquid saturated sample when drying starts. As a result of evaporation the gas phase progressively invades the material leading to the formation of a two-phase zone within the material where pores are either occupied by the liquid or the gas phase (the possible presence of liquid films in the pores occupied by the gas phase is not considered for the moment). Generally, the development of the two-phase zone in a capillary porous medium under the quasi-isothermal condition considered in this chapter has little impact on the evaporation rate, at least during a first phase of drying referenced to as the constant rate period (CRP). This is qualitatively illustrated in Fig. 7.3. The remarkable result is that the overall liquid saturation diminishes without a noticeable change in the evaporation rate during the CRP. This has to do with the phenomenology described in Sect. 7.2. Figure 7.5 shows an example of phase distribution at the surface during the CRP obtained from PN simulations,

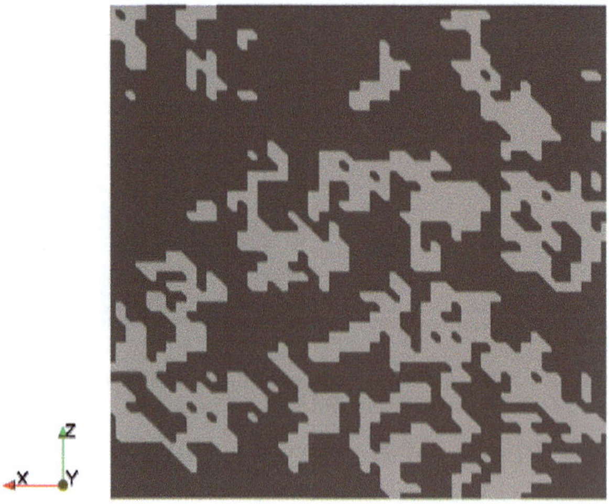


**Fig. 7.5** Liquid and gas distribution at the surface of a drying porous medium from pore network simulations. Gas patches in *light grey*. Liquid in *dark grey*

Lebray and Prat (1999). The evaporation rate does not change noticeably as long as the average size of gas/dry patches is small compared to the mass boundary layer thickness. There is a compensation mechanism. The decrease in the evaporation flux from "dry" pores at the surface is compensated by the increase in the evaporation flux from the pores at the surface remaining occupied by the liquid phase. Note that the influence of film effects on the duration of constant rate period is discussed in Sect. 7.4.2. It can then be understood, at least qualitatively, that the aforementioned compensation mechanism will become less effective when the extent of dry region at the surface reaches a sufficiently large surface area. The consequence is a decrease in the evaporation rate. This corresponds to the falling rate period (FRP). Liquid is still present at the surface but not sufficiently to maintain a constant evaporation rate. Eventually all the liquid disappears at the surface (not considering the very tiny liquid bridges which may still be present) and this corresponds to the receding front period (RFP) characterized by the development of a dry zone adjacent to the open surface of porous medium. Evaporation takes place inside the material during the RFP.

A classification of drying regimes can then be obtained from the maximum extent of two-phase zone during the drying process. The CRP is expected to last until the surface ceases to be hydraulically connected to the liquid located further inside the porous sample. Thus moment, i.e. the end of the CRP should roughly correspond to the moment when the two phase zone reaches its maximum extent (under the additional condition that the two-phase zone is still attached to the surface). This is illustrated in Fig. 7.6. The maximum extent $L$ of two-phase zone can be in fact regarded as a characteristic length, e.g. Lehmann et al. (2008). The end of CRP roughly corresponds to the moment where the gas phase reaches the depth $L$ within the material. For the configuration depicted in Fig. 7.1a, gravity and viscous forces are stabilizing, see Prat and Bouleux (1999) or Prat (2002) for more details. This means here that both forces tend to limit the extent of two-phase zone. Simple but quite rough estimates of characteristic length $L$ can be

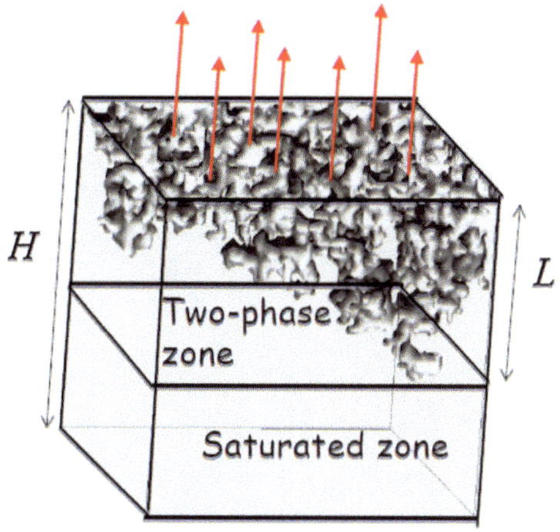


**Fig. 7.6** Concept of characteristic length for analyzing the drying process

obtained from classical properties of porous material such as the retention curve, the permeability and the relative permeabilities, see Lehmann et al. (2008). Improved determinations have been proposed, e.g. Assouline et al. 2014, but the concept remains more qualitative than quantitative because notably of the film effects discussed in Sect. 7.4.2.

In fact gravity and viscous effects rarely are significant together. Roughly the characteristic length results from the equilibrium between gravity and capillary effects across the two-phase zone in sufficiently coarse porous media. By contrast, gravity effects are negligible compared to viscous effects in sufficiently fine porous media, where the characteristic length then results from the equilibrium between viscous and capillary effects across the two-phase zone. In brief, we can define a capillary-gravity characteristic length $L_g$ and a capillary—viscous characteristic length $L_v$. The smallest of these two lengths characterizes the extent of two-phase zone: hence $L \approx \min(L_g, L_v)$. This eventually leads to distinguish three main drying regimes depending on $L_v$, $L_g$ and the sample height $H$. The capillary—viscous and capillary-gravity drying regimes are obtained when $L \ll H$. The former when $L = L_v \ll H$ and the latter when $L = L_g \ll H$. Both are characterized by a travelling two-phase zone within the material. The capillary regime is obtained when $L \gg H$. In this case the two phase zone spans the whole height of sample over most of CRP. This regime is frequently encountered in standard experiments with building materials at room temperature, e.g. Gupta et al. (2014). This can be understood from the computation of characteristic lengths. Consider for simplicity random packings of monodisperse particles of diameter $d$. As reported in Hidri et al. (2013), the retention curve $P_c(S)$ of such a system can be described using a standard van Genuchten representation as

$$\frac{S - S_c}{1 - S_c} = \frac{1}{\left[1 + \left(\frac{P_c}{P_{cref}}\right)^n\right]^m} \tag{7.4}$$

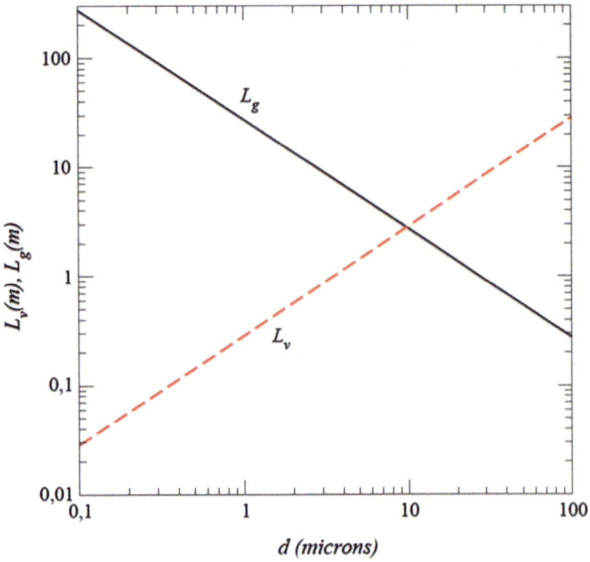


**Fig. 7.7** Variation of characteristic lengths as a function of bead diameter for random packings of monodisperse particles

with $m = 1 - \frac{1}{n}$ and $P_{cref} = \frac{6(1-\varepsilon)\gamma}{\varepsilon d}$, where $\gamma$ is the surface tension ($\gamma = 72 \times 10^{-3}$ N/m for pure water), $n = 10$, $S_c = 0.1$ and $\varepsilon$ is the porosity $\varepsilon \approx 0.4$. According to Lehmann et al. (2008) estimate of $L_g$ and $L_v$ are then given by,

$$L_g \approx \frac{P_{cref}}{\rho_\ell g(n-1)} \left(\frac{2n-1}{n}\right)^{(2n-1)/n} \left(\frac{n-1}{n}\right)^{(1-n)/n} \tag{7.5}$$

and

$$L_v = \frac{\rho_\ell g k k_r(S)}{\mu\, e_0} L_g \tag{7.6}$$

where $\rho_\ell$ is the liquid density, $\mu$ is the dynamic viscosity of the liquid and $e_0$ the evaporation flux during the CRP expressed in m/s ($e_0 = J/A/\rho_\ell$ where $J$ is the evaporation rate (kg/s) and $A$ is the top surface area of porous sample). The permeability in Eq. (7.6) can be estimated using the Carman–Kozeny relationship $k = \frac{\varepsilon^3 d^2}{180(1-\varepsilon)^2}$ and the liquid phase relative permeability as $k_r(S) = \frac{(S-S_c)^3}{(1-S_c)^3}$. As representative value of evaporation flux under usual laboratory conditions, we can take $e_0 = 1$ cm/day. To estimate the relative permeability we take an average value of saturation, i.e. $S = 0.5\ (1 + S_c)$. For water, this leads to the results plotted in Fig. 7.7.

In typical drying experiments at the laboratory scale, the height $H$ of sample is generally on the order of a few cm. Capillary porous media corresponds to pore sizes greater than about 100 nm, which corresponds to bead size greater than about 1 μm. As can be seen from Fig. 7.7, $L_v$ and $L_g$ are then much greater than $H$, i.e. say 5 cm,

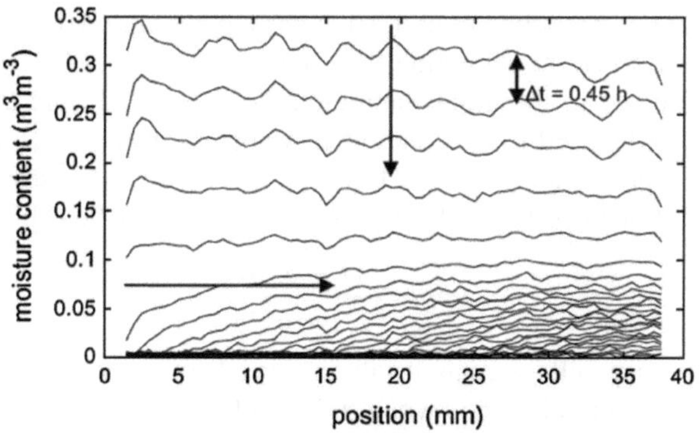


**Fig. 7.8** Measured moisture profiles during drying of a fired-clay brick sample. The profiles are given for every 0.45 h. The drying surface is at 0 mm. The *vertical arrow* shows the homogenous drying of the sample (externally limited) and the *horizontal arrow* shows the penetration of the receding drying front (internally limited). Reprinted from (Gupta et al. 2014) with permission from Elsevier

and the prevailing regime is the capillary regime. The type of calculation leading to the results plotted in Fig. 7.7 can be repeated for capillary building materials, i.e. fired-clay brick for instance, to determine the prevailing regime under laboratory or building/field conditions. This notably allows making simplifications in the modelling depending on the scale hierarchy between $H$, $L_v$ and $L_g$. For example, as shown in Hidri et al. (2013), the evolution of saturation profiles during the CRP can be determined using only a simple mass balance and the retention curve when $L_v \gg \max(H, L_g)$.

Since the capillary regime is frequently encountered in typical laboratory conditions, additional features are presented. The most striking feature is that the saturation can be essentially spatially uniform along the sample during a first part of CRP. This is illustrated in Fig. 7.8 in the case of the drying of a fired-clay brick sample.

The evolution of overall mean saturation can be predicted from the simple mass balance during the CRP,

$$S = 1 - \frac{e_0}{H\varepsilon} t \tag{7.7}$$

In the capillary regime $S$ is also the saturation all along the sample as long as the saturation is spatially uniform.

The next question is predicting when the CRP ends. It is customary to consider that the CRP ends when the saturation reaches the "critical" saturation. The concept of critical saturation is however only approximate. In fact, as illustrated in Fig. 7.8, the saturation profile is not spatially uniform all along the CRP. It tends to bend as the end of CRP is approached as a consequence of viscous effects. Hence a more accurate description of capillary regime is to consider that viscous effects do not affect the phase distribution over most of CRP but become non negligible toward the end of CRP as the liquid phase becomes increasingly poorly connected. This leads to consider that the average saturation marking the end of CRP depends in fact of evaporation rate, see Assouline et al. (2014) or Gupta et al. (2014) for more details. However, when $L_v \gg H$, the saturation $S_{CRP}$ marking the end of CRP can be expected to be small compared to the initial saturation, i.e. $S_{CRP} \ll 1$, and close to irreducible saturation $S_c$ (deduced from retention curve determination). As a result, the end of CRP can be approximately derived from Eq. (7.7) with $S_c$ as "critical" saturation. This yields the following estimate for the duration of CRP in the capillary regime,

$$t_{CRP} = \frac{(1 - S_c)H\varepsilon}{e_0} \tag{7.8}$$

Hence in the pure capillary regime, also referred to as the asymptotic capillary regime, the CRP essentially ends when the liquid phase is in a pendular stage, i.e. is formed by a set of disconnected clusters, see Le Bray and Prat (1999) for more details. The next phase, i.e. the FRP/RFP, can then be modelled using a simple receding front model, considering that the saturation is everywhere equal to $S_c$ at the end of CRP. This leads to this simple expression for estimating the total drying time $t_F$ (the details will be presented in a forthcoming paper)

$$\frac{t_F}{t_{CRP}} = \frac{\left(0.5H + \frac{\varepsilon D_{eff}}{D}\,\delta\right)S_c}{\frac{\varepsilon D_{eff}}{D}(1 - S_c)\delta} + 1 \tag{7.9}$$

where $D_{eff}$ is the effective coefficient of vapour diffusion in the porous medium. In the case of a packing of particles, $\frac{\varepsilon D_{eff}}{D} \approx \varepsilon^{1.4}$, e.g. Kim et al. (1987) and Eq. (7.9) can be further simplified as,

$$\frac{t_F}{t_{CRP}} = \frac{\left(0.5H + \varepsilon^{1.4}\,\delta\right)S_c}{\varepsilon^{1.4}(1 - S_c)\delta} + 1 \tag{7.10}$$

This quite simple model of the capillary regime which only requires to know the porosity of the medium, the "critical" saturation $S_c$, the height of sample and the external mass transfer characteristic length $\delta$ (or equivalently the potential evaporation) was tested against the experimental data for a packing of beads 300 μm in diameter reported in Eloukabi et al. (2013). It was found that the predictions given by Eqs. (7.8) and (7.10) were in good agreement with the experiment. Both the duration of CRP and final drying tine were well predicted. The details will be soon presented elsewhere.

Again, it should be clear that the capillary regime considered above is asymptotic in the sense that viscous effects in the liquid saturated pores are assumed to be negligible compared to capillary effects over all the CRP. In this regime, the saturation profile remains flat until the end of CRP. This is the expected regime when $H \ll L = \min(L_g, L_v)$. When $H$ is not sufficiently small compared to $L$, the profile is not flat anymore and numerical solutions can be necessary to predict the drying process, especially in the case where $L = L_v$. This will be illustrated in Sect. 7.6. Also, we have essentially considered a 1D problem. Problems where evaporation takes place from several sides of porous domain can be also handled analytically up to the end of CRP in the asymptotic capillary regime but the prediction of final drying time (through the modelling of receding front period) generally requires a numerical approach.

The predictions proposed by the very simple model presented in this section are implicitly based on the assumption that liquid film effect is negligible. As discussed in the next section, the modelling of the capillary regime (as well as the two other regimes) is completely to reconsider when film effects can be expected to be significant (evaporation of a perfectly wetting liquid for instance).

### *7.4.3 Influence of Liquid Films*

One can distinguish two types of liquid films: the capillary films and the adsorbed films. Adsorbed films are very thin films of a few nanometers in thickness. They are classically invoked as a possible form of bound water in hygroscopic material. Capillary films are much thicker films which develop a long the corners wedges, crevices, etc. of the pore space. Figure 7.9 shows an example of capillary films in a model system made of a monolayer of glass beads sandwiched between two transparent plates. The region shown in Fig. 7.9 has been invaded by the gas phase as a result of evaporation. Capillary liquid films partially carpet the beads and maintain the hydraulic connectivity in this region where most of the pores are invaded by the gas only in the bulk.

The impact of capillary films on drying kinetics can be huge when the wettability and geometrical conditions are favourable for the development of the films over a long distance. As reported in Chauvet et al. (2009), the overall drying time of a square tube can be orders of magnitude faster than a circular capillary tube of same aperture and length because of the effect of the corner capillary films in the square tube. Spectacular capillary liquid film effects were also reported for porous systems formed by beads confined in narrow slits, Yiotis et al. (2012). A qualitative explanation is given in Fig. 7.10. In the presence of liquid films, the evaporation front is, for approximately the same overall saturation, much closer to the open surface of porous media than when the film effect is negligible. Since the evaporation rate is roughly inversely proportional to this distance, the evaporation rate is significantly greater for the situation depicted in Fig. 7.10a compared to the situation in Fig. 7.10b. Also the films can be attached to the porous medium surface long after the liquid bulk connectivity to the surface has ceased. The result is a CRP much longer than predicted from the consideration of the bulk hydraulic connectivity only.

Although there are spectacular experimental evidences of the key role that the liquid films can play on drying in model systems, the situation is much less clear for real porous media and in particular for building materials. Some indirect evidences have been reported from the analysis of drying from soil columns, e.g. Assouline et al. (2014) but the quality of the data were not sufficient for reaching a firm conclusion on this point. In packing of particles, lace like film structures have been identified, Wang et al. (2013), that is a 3D version of the type of capillary structure shown in Fig. 7.9. The role plays by this structure in the drying of 3D packings of particles is still unclear, e.g. Wang et al. (2012) . This is in fact an interesting open

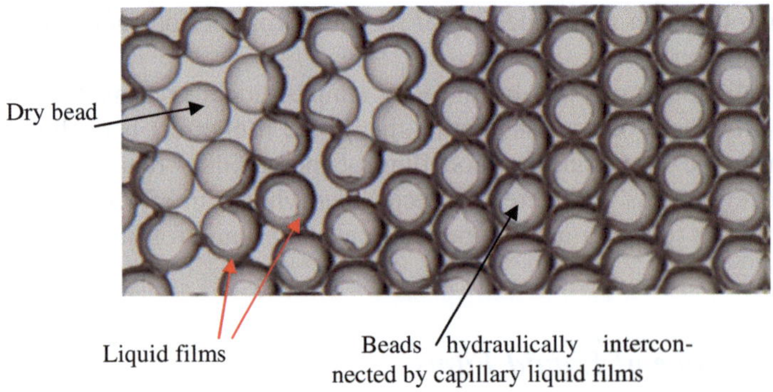


**Fig. 7.9** Capillary liquid films in a model porous medium

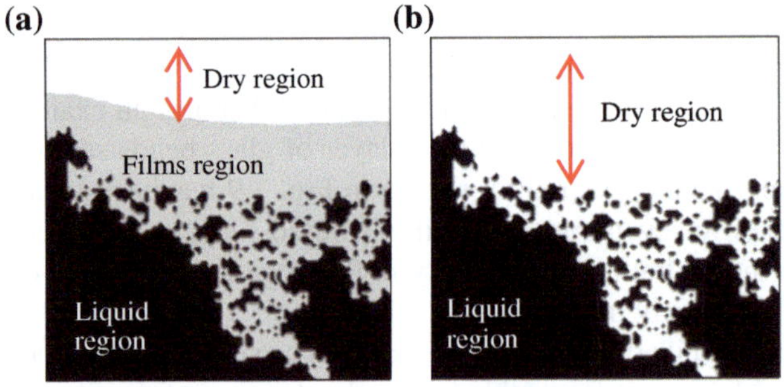


**Fig. 7.10** Schematic of film effects. The *red arrow* schematically characterizes the distance between the porous medium surface and the internal evaporation front. The greater the *arrow*, the lower the drying rate

area of research and perhaps the currently most important question regarding the physics of drying porous media.

## 7.5 Physics of Drying of Hygroscopic Porous Media

A major difference between capillary porous media and hygroscopic porous media lies in the fact that the equilibrium vapour pressure at the porous medium surface depends on the saturation at the surface, denoted by $S(0)$, even in the first phase of drying where the transport of liquid to the surface due to capillary action is sufficient to equilibrate the external evaporative demand. Physically, this is consequence of the fact that the equilibrium vapour pressure not only depends on the temperature but also on the curvature of the liquid-gas interface when the radius of curvature is sufficiently small. Thus Eq. (7.3) during this period reads

$$j_0 = D\frac{M_v}{RT}\frac{\left(p_{veq}(S(0)) - p_{v\infty}\right)}{\delta} \tag{7.11}$$

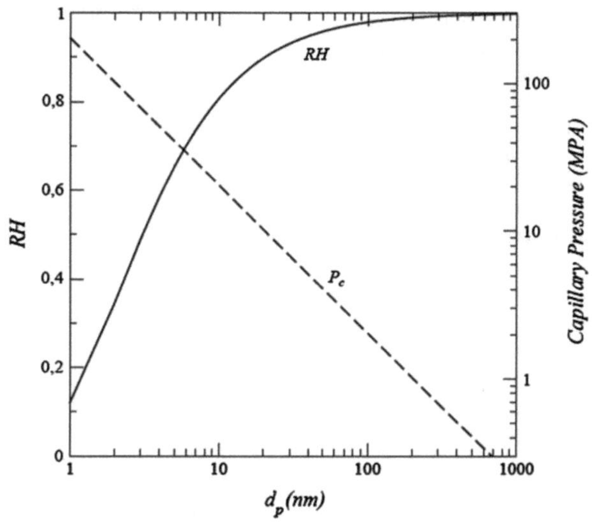


**Fig. 7.11** Equilibrium relative humidity ($RH = p_{veq}/p_{vs}$) as a function of pore diameter $d_p$ according to Kelvin's relationship for water and capillary pressure (computed assuming a *cylindrical* pore)

where $p_{veq}(S(0))$ is generally determined using the Kelvin-Laplace equation as,

$$\frac{p_{veq}}{p_{vs}} = \exp\left(-\frac{M}{RT}\frac{P_c(S(0))}{\rho_\ell}\right) \quad (7.12)$$

where $p_{vs}$ is the saturation vapour pressure (= the equilibrium vapour pressure over a flat liquid—gas interface). It is generally considered that Kelvin-Laplace equation is valid for equilibrium relative humidity greater than about 50 %. According to Fig. 7.11, this corresponds to pore sizes of about 3–4 nm. Also Fig. 7.11 indicates that the Kelvin-effect is significant for pore sizes below a few hundred nanometers, consistently with the "critical" pore size separating the hygroscopic and capillary porous materials.

As a result we can in fact distinguish different situation depending on the pore size distribution focussing again only on the first period of drying. If all pores are greater than 1 μm, a CRP is expected since the water activity, i.e. the equilibrium vapour pressure, is independent of menisci curvature at the surface. This corresponds to the capillary porous media. By contrast, no CRP is expected, even at extremely low drying rates, when the medium is purely hygroscopic, i.e. all pores are lower than, say, 100 nm. This is illustrated in Fig. 7.3. For broader pore size distribution with significant fraction of pores in both the submicronic range and above 1 μm, then a CRP can be observed followed by a decrease in the drying rate when the pore size containing liquid at the surface are all in the submicromic range. This decrease is different from the one observed with capillary regime in the FRP since it is not due to the impact of the increase in the surface fraction occupied by dry pores but to the Kelvin's effect. Naturally at later stages, both effects can contribute to the decrease in the drying rate.

A major difference between capillary porous and hygroscopic porous media is that drying can end for an intermediate value of saturation with a hygroscopic

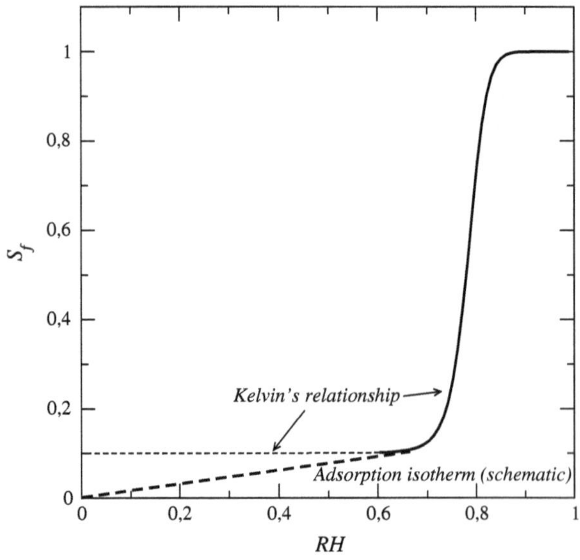


**Fig. 7.12** Saturation at the end of drying as a function a relative humidity in the external air for a hygroscopic material (corresponding to a random packing of beads 20 nm in diameter) as predicted from the Kelvin-Laplace equation

porous medium whereas full drying is always expected for a capillary porous medium. The equilibrium saturation, i.e. the saturation marking at the end of drying; is obtained by solving Eq. (7.13). As an example consider a packing of monodisperse spherical beads 20 nm in diameter. Solution of Eq. (7.13) using Eq. (7.4) for the retention curve then leads to the results shown in Fig. 7.12.

Figure 7.12 of course corresponds to the part of the desorption isotherm of the material that can be described thanks to the Kelvin-Laplace equation.

The continuum model classically used for describing the drying at moderate (room) temperature of hygroscopic materials is similar to the one for capillary porous media (see Sect. 7.6). Thus the physics of the pore scale involves the mechanisms listed in Sect. 7.4.1. The gravity effects are, however, negligible for the sample sizes generally considered in laboratory experiments. Also, in addition to desorption phenomena, it can be necessary to take into account rarefaction effects when the mean free path of molecules in the gas phase is not small compared to the pore size (we recall that the mean free path is on the order of 70 nm at the atmospheric pressure and room temperature, thus greater than the sizes of smaller pores in hygroscopic materials).

For drying of samples submitted to high heat fluxes, the total gas pressure cannot be assumed as a constant, e.g. Perré et al. (2007). By contrast, it is generally assumed that the total gas pressure can be assumed as constant, equal to the atmospheric pressure and spatially uniform during (slow) quasi-isothermal drying at room temperature However, this assumption was questioned by Mainguy et al. (2001) in their analysis of drying of concrete. Their modelling led to the conclusion that a significant total pressure variation occurred during drying, even under the slow drying conditions (=drying at room temperature). This question was reconsidered again in Thiery et al. (2007). Questioning the permeability choice made by Mainguy et al. (2001), their results were in favour of the usual assumption of a negligible total pressure effect on the drying process. Actually, this issue is perhaps not completely clarified and the fact that the total pressure could play a role in the drying process of hygroscopic materials even under slow drying condition can be kept in mind when analyzing modelling and experiments which do not match well.

As discussed in Sect. 7.4.3, the capillary liquid films are a key issue in the analysis of the drying of capillary porous materials. Their role in the drying of hygroscopic porous materials is less clear and needs to be assessed. Somewhat similarly as for capillary systems, it has been shown, however, that they can have a major impact on the drying kinetics of a model hygroscopic system (Eijkel et al. 2005).

## 7.6 Continuum Approach to Drying

Darcy's scale models, where the porous medium is represented as a fictitious continuum, are the most classical ones and certainly the most used in the applications for modelling the drying process. The corresponding set of equations can be obtained directly at the level of the fictitious continuum medium using phenomenological approaches, e.g. Philip and de Vries (1957), Luikov (1966) or using more rigorous frameworks, such as the thermodynamic of irreversible processes, Benet and Jouanna (1983) or the theory of mixtures (Kowalsky 2007). The corresponding equations can also be obtained using upscaling techniques, such the volume averaging method, e.g. Whitaker et al. (1977), Whitaker (1998). As mentioned earlier, we considered in this chapter the relatively simple situation where the porous microstructure is non-deformable, temperature gradients are negligible, and the pressure in the gas phase is uniform and constant, equal to the atmospheric pressure. The reader interested in more complex situations can refer for example to (Perré et al. 2007) and references therein. For our conditions, the continuum model equations can be listed as follows, beginning with the mass conservation equations which read

$$\frac{\partial[\rho_v \varepsilon(1-S)]}{\partial t} + \nabla \cdot J_v = \dot{m} \tag{7.13}$$

$$\frac{\partial[\rho_\ell \varepsilon S]}{\partial t} + \nabla \cdot J_w = -\dot{m} \tag{7.14}$$

where $\rho_\ell$, $\rho_v$, $\varepsilon$, $S$, $t$ are the liquid density, water vapour density, porous material porosity, water saturation and time respectively; $\dot{m}$ is the evaporation rate per unit volume of porous medium (kg/m$^3$/s); $J_v$ is the water vapour mass flux whereas $J_w$ is the liquid water mass flux. It is assumed that the water vapour and air binary gas mixture behaves as a perfect gas

$$\rho_v = \frac{p_v M_v}{RT} \tag{7.15}$$

The liquid water flux in Eq. (7.13) is expressed using the generalized Darcy's law as

$$J_w = -\rho_\ell \frac{k\,k_{r\ell}}{\mu}(\nabla p_\ell - \rho_\ell \mathbf{g}) \tag{7.16}$$

where $p_\ell$ the liquid pressure and **g** the gravity vector.

The vapour flux in Eq. (7.13) is expressed using Fick's law as

$$J_v = -\rho_\ell \frac{M_v}{RT}\frac{p_{atm}}{p_{atm} - p_v} f(S) D \nabla p_v \tag{7.17}$$

where $f(S)$ is a function of saturation introduced for taking into account the fact the diffusion of water vapour in a partially saturated porous medium is different from diffusion in a free gas mixture. $f(S)$ can be regarded as one of the weakest points of the continuum model since there is no widely accepted approach to characterize it. One option is to consider that liquid water reduces the passages left free for the vapour transport. Accordingly this leads some authors to propose expressions of the form,

$$f(S) = \varepsilon^a (1 - S)^b \tag{7.18}$$

where $a$ and $b$ are positive exponents. The exponent $a$ can be specified so as to be consistent with diffusion transport in a dry material, i.e. $a \approx 4/3$ for a granular material for instance. The exponent $b$ can be determined considering the diffusion of a gas different from water vapour, i.e. not interacting with the water clusters present in the medium. As an example, this leads to $b = 10/3$ for a granular material, Millington (1959). However, the situation can be different when water vapour diffuses in a partially saturated porous medium because of evaporation-condensation phenomena occurring at the liquid-gas interfaces. According to Philip and de Vries (1957) the net result is short circuit effect leading to a possible enhancement of the vapour transport. The effect can be taken into account by introducing the enhancement factor $\eta(S)$,

$$f(S) = \eta(S)\varepsilon^a (1 - S)^b \tag{7.19}$$

The enhancement can be quite significant (maximum values as high as four or greater have been reported), e.g. Plumb (2000) and references therein. This aspect of the model still needs further investigations.

Invoking the local capillary equilibrium assumption, the liquid pressure is expressed as,

$$p_c(S) = p_{atm} - p_\ell \tag{7.20}$$

where $p_c(S)$ is the retention curve of the porous material.

A simple one equation model is obtained by adding Eq. (7.13) and Eq. (7.14) taking into account the above equations,

$$\frac{M_v}{RT}\frac{\partial[p_v \varepsilon(1 - S)]}{\partial t} + \frac{\partial[\rho_\ell \varepsilon S]}{\partial t} = \nabla.\left[\rho_\ell \frac{K\,k_{r\ell}}{\mu}\left(-\frac{\partial p_c}{\partial S}\nabla S - \rho_\ell g\right) + \frac{M_v}{RT}\frac{p_{atm}}{p_{atm} - p_v} f(S) D \nabla p_v\right] \tag{7.21}$$

Here we have chosen $S$ as main variable. Naturally other choices are possible and can be more adapted to a given situation, e.g. (Plumb 2000). As reported in (Hidri et al. 2013), it is better for example to use the liquid pressure as main variable when dealing with heterogeneous porous media (layered porous media for example) or situations where saturated and unsaturated regions coexist in the system.

Equation (7.21) can be further simplified by invoking that the transfer of the vapour is quasi-steady. This leads to

$$\frac{\partial[\rho_\ell \varepsilon S]}{\partial t} = \nabla \cdot \left[\rho_\ell \frac{K\, k_{r\ell}}{\mu}\left(-\frac{\partial p_c}{\partial S}\nabla S - \rho_\ell \mathrm{g}\right) + \frac{M_v}{RT}\frac{p_{atm}}{p_{atm}-p_v} f(S) D \nabla p_v\right] \tag{7.22}$$

The next step is to find a relationship between $S$ and $p_v$,

$$\frac{p_v}{p_{vs}} = \phi(S) \tag{7.23}$$

Equation (7.23) corresponds to a desorption isotherm for a hygroscopic material. As mentioned in Sect. 7.3, sorption phenomena can be neglected in capillary porous media. The introduction of Eq. (7.23) should then be considered as a purely numerical trick enabling one to define a liquid saturation even in the completely dry region of the sample (during the receding front period). This avoids tracking explicitly the interface between the two-phase zone and the dry region.

Combining Eqs. (7.22) and (7.23) eventually leads to

$$\frac{\partial[\rho_\ell \varepsilon S]}{\partial t} = \nabla \cdot \left[\rho_\ell \frac{K\, k_{r\ell}}{\mu}\left(-\frac{\partial p_c}{\partial S}\nabla S - \rho_\ell \mathrm{g}\right) + \frac{M_v}{RT}\frac{p_{atm}}{p_{atm}-p_v} f(S) D p_{vs} \frac{\partial \phi}{\partial S}\nabla S\right] \tag{7.24}$$

To solve the above equation boundary and initial conditions must be supplemented. As an example consider the drying situation depicted in Fig. 7.1a. At the surface in contact with air we express the continuity of the mass flux as

$$\begin{aligned}&\left[\rho_\ell \frac{K k_{r\ell}}{\mu}\left(-\frac{\partial p_c}{\partial S}\nabla S - \rho_\ell \mathrm{g}\right) + \frac{M_v}{RT}\frac{p_{atm}}{p_{atm}-p_v} f(S) D p_{vs} \frac{\partial \phi}{\partial S}\nabla S\right]\cdot \mathbf{n} \\ &= \frac{M_v}{RT} D\left(\frac{p_v(0) - p_{v\infty}}{\delta}\right)\end{aligned} \tag{7.25}$$

where $p_v(0)$ is the vapour pressure at the surface.

The zero flux condition at the bottom reads,

$$\left[\rho_\ell \frac{K\, k_{r\ell}}{\mu}\left(-\frac{\partial p_c}{\partial S}\nabla S - \rho_\ell \mathrm{g}\right) + \frac{M_v}{RT}\frac{p_{atm}}{p_{atm}-p_v} f(S) D p_{vs} \frac{\partial \phi}{\partial S}\nabla S\right]\cdot \mathbf{n} = 0 \tag{7.26}$$

Many works have been devoted to test this model against experimental data, especially in the eighties, e.g. Chen and Pei (1989), and references therein. However, the "good" agreement generally found by the various authors is rarely fully convincing because of the frequent use of adjustable parameters, such as for example the external mass transfer coefficient [which is equivalent to the length $\delta$ in Eq. (7.3)], which is made depending on the surface saturation in an ad-hoc way. Although leading to results qualitatively in agreement with the experiments, e.g. Coussot (2000), it seems that something is still missing in this type of model for obtaining truly predictive results. This might be due to the liquid film effects, which is difficult to take into account explicitly in this type of model or to still insufficient analysis of the transfer at the evaporative surface of the porous medium, especially toward the end of the constant rate period and during the falling rate period. As any model, the continuum model thus represents an approximation, which can be fully satisfactory or not depending on the objective. Despite the remaining questions just mentioned, this model can be quite useful, especially for engineering applications.

As an example we use here this model to predict the influence of pore size on the duration of CRP in the capillary regime when the viscous effects are not negligible up to the end of CRP. We recall that the capillary regime is defined as the regime where the two-phase zone can span the sample (as opposed to the situation where the two-phase zone maximum extent is lower than the height of sample). We discussed in Sect. 7.4.2, the case where the viscous effects are negligible up to the end of CRP. The final saturation was then $S_c$ and the duration of CRP is given simply by Eq. (7.8). As noted in Sect. 7.4.2 this represents the asymptotic capillary regime, where viscous effects are totally neglected. To determine the range of validity of asymptotic capillary regime, we neglect the gravity effects and assume that the transport of vapour by diffusion is negligible during the CRP. Thus we solve the equation

$$\frac{\partial[\rho_\ell \varepsilon S]}{\partial t} = \nabla \cdot \left[\rho_\ell \frac{K\, k_{r\ell}}{\mu}\left(-\frac{\partial p_c}{\partial S}\right)\nabla S\right] \tag{7.27}$$

together with a zero flux condition at the impervious side o sample and the condition

$$\left[\rho_\ell \frac{K\, k_{r\ell}}{\mu}\left(-\frac{\partial p_c}{\partial S}\right)\nabla S\right] \cdot \mathbf{n} = j_0 \tag{7.28}$$

at the evaporative surface.

Then we determine the end of CRP as the moment when Eq. (7.28) cannot be satisfied. Numerically, this means that we consider that the CRP ends when the liquid flux (l.h.s. in Eq. (7.28)) predicted from the saturation profile obtained at time $t$ is equal to 98 % $j_0$. This gives the results plotted in Figs. 7.13 and 7.14. As can be seen from Fig. 7.13, the time marking the end of CRP decreases with a decreasing bead size, in qualitative agreement with the results reported for example in Coussot (2000). The inset in Fig. 7.13 shows the computed saturation at the surface at the end of CRP. As can be seen, this saturation is close to but greater than $S_c$ and varies with the bead size.

Figure 7.14 illustrates the fact that the saturation profile is increasingly less flat as the impact of viscous effects increases.

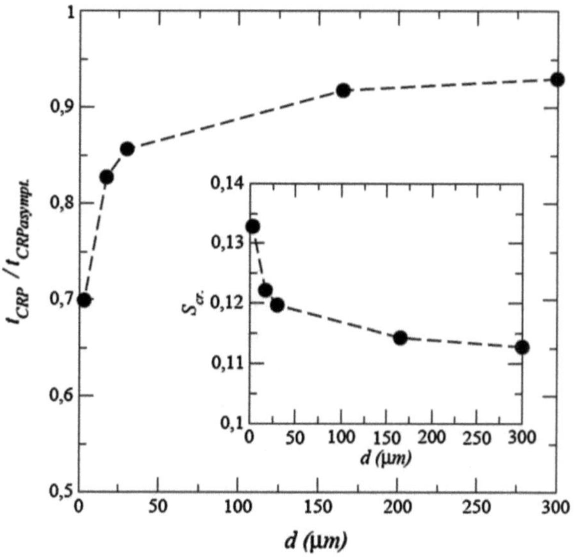


**Fig. 7.13** Time marking the end of CRP for various bead sizes. The reference time $t_{CRPasympt}$ is the time for the asymptotic capillary regime given by Eq. (7.8). The *inset* shows the variation of "critical" saturation as a function of bead size. The critical saturation is the saturation at the porous medium surface at the end of CRP

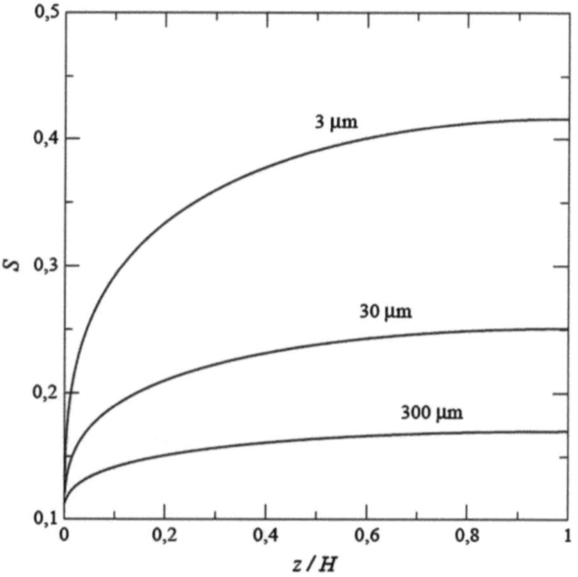


**Fig. 7.14** Computed saturation profiles at the end of CRP for various bead sizes

## 7.7 Advanced Modelling Methods

### *7.7.1 Overview*

As in other fields dealing with porous media, the modeling of drying phenomena can be performed at various scales using different techniques. Although mainly motivated by wood drying, an overview of the field is given by Perré (2011).

At the smallest length scales and time scales, molecular simulations can be performed as a tool to understand the physics at the scale of a single pore. At a slightly greater scale, simulations of a porous material as aggregates of particles can be performed, e.g. Masoero et al. (2012). It seems that these techniques have not been used very much in relation with drying problems in porous media so far. At the scale of a few pores, direct simulations of the transport processes refer to simulations where the pore scale continuum equations, e.g. the Navier-Stokes equations for example, are solved. Simplified direct simulations of the drying process under quasi-static conditions in a random packing of particles using the volume of fluid (VOF) method are presented in Kohout et al. (2006). Equally, although restricted to quasi-static phase distributions, the works of Bentz et al. (1995, 1998) on the drying of hygroscopic materials can also be considered as direct simulations. Although direct simulations have scarcely been used so far in relation with drying, it can be expected that this situation will change rapidly, notably as regards the dynamic simulations. The next scale when describing models from the pore scale to the Darcy's scale is the pore network scale. This scale means simulations of the drying process over domains containing typically $10^5$–$10^6$ pores. This corresponds to the pore-network models, a modelling technique which has been the subject of quite a few works in relation with the study of drying of capillary porous media. Pore network models are discussed in the next section.

## *7.7.2 Pore Network Models*

The study of drying from pore network simulations has become relatively popular, see the review papers Prat (2002) and (2011) and references therein, Pore network models are well adapted to represent "simple" porous materials characterized by a relatively narrow pore size distribution. Bricks and some stones are reasonable examples of this type of materials. We will refer to this type of pore network model as one scale pore network models. Note however that multiscale pore network models have been developed for the study of materials characterized by broad pore size distributions, e.g. Xu et al. (1997a, b), Carmeliet et al. (1999), but mostly as a tool for determining macroscopic properties and not for simulating directly the drying process. Hence, one can actually distinguish two main uses of pore network models. They can be used at the scale of the representative elementary volume for determining the macroscopic parameters of the Darcy's scale continuum models such as, the permeability, the relative permeabilities, the effective diffusion coefficient or the capillary pressure curve, e.g. Xu et al. (1997a, b), Blunt et al. (2002). Pore network models can be also used directly as a tool to study the drying process regardless of the continuum modelling. Since there exist recent reviews, e.g. Metzger et al. (2007), Prat (2011), the objective in this chapter is not to present again a review.

We simply give an introductory presentation to the technique and some comments. The interest of the technique is illustrated through the consideration of an

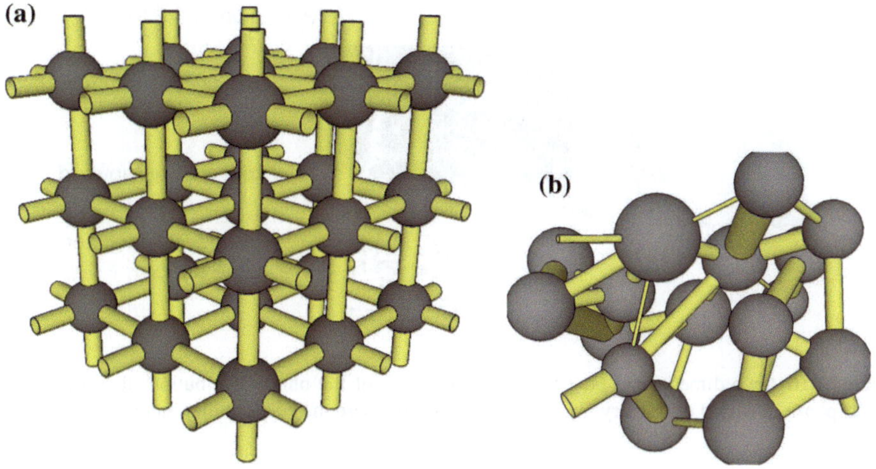


**Fig. 7.15** Structured **a** and unstructured **b** pore networks

example. Pore network models are based on a conceptualization of the pore space as a network of pores connected by narrower channels. The pores are located at the nodes of the networks. The interconnecting channels between pores correspond to the constrictions of the pore space. As sketched in Fig. 7.15, the pore network model can be structured when the pores are located at the nodes of a regular grid (a cubic grid in Fig. 7.15a) or unstructured (Fig. 7.15b) when the pore network is constructed for example from a digital image of a real porous microstructure (see Sect. 7.9).

Once the pore space is partitioned into pores and throats, a drying algorithm can be implemented. The simplest drying algorithm, Prat (1993), combined the invasion percolation algorithm applied to each liquid cluster forming during drying to a finite-volume like steady-state solution of the vapour diffusion problem in the gas phase. Since only capillary effects are taken into account as regards the computation of the pressure field in the liquid, this drying algorithm is adapted to simulate very slow drying conditions. Naturally, many other phenomena have been taken into account since this first algorithm was proposed, e.g. Prat (2002, 2011). However, it can be pointed out that the available dynamic pore network models of drying are for capillary porous media. Extension to hygroscopic porous materials is in progress.

The explicit consideration of liquid film flows, see Sect. 7.4.3, is perhaps the most remarkable contribution of pore network models to our understanding of the physics of drying. However, as mentioned in Sect. 7.4.3, a great deal of work is still needed to develop more representative models of liquid film flows in real materials than the currently available model, which represents the pore space in the film region as a network of interconnected channels of simple polygonal cross section.

Another interesting result is the contribution of pore network simulations to the understanding of the localisation of salt crystallization spots at the evaporative surface of porous medium, see Sect. 7.8.

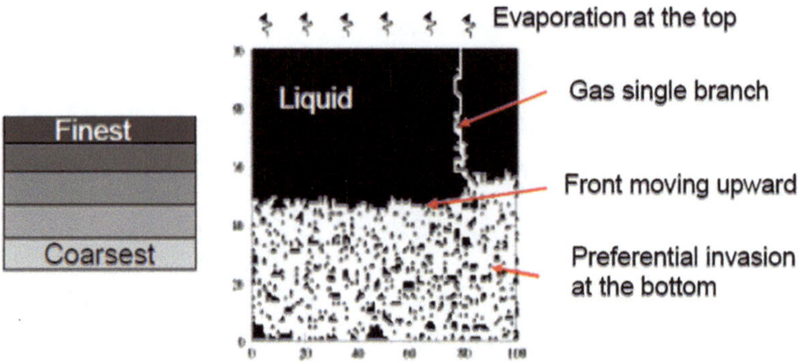


**Fig. 7.16** Two-dimensional pore network simulation of the phase distribution (liquid in *black*, gas phase in *white*) during drying in a porous system where the permeability increases *downward*

Pore network simulations are also interesting to study the impact of large scale heterogeneities such as permeability heterogeneities as briefly illustrated in what follows.

Consider a porous medium where the local permeability increases from the open evaporative surface to the bottom as sketched in Fig. 7.16. This corresponds to the case of a layered porous system for example.

Again we discuss only slow drying conditions so that the invasion of the system by the gas phase due to evaporation is dominated by capillary effects. As illustrated in Fig. 7.16, drying in the case is first characterized by the development of a thin single gas branch developing from the evaporative surface up to the bottom of system. The next phase of the invasion is characterized by the preferential invasion of the region adjacent to the bottom. Thus, evaporation takes place at the top but invasion takes place far away from the top over a significant period of drying. The fact that the evaporative surface remains highly saturated over a long period favours a long constant rate period. More details on this drying situation will be given in a forthcoming paper.

## 7.8 Ion Transport, Crystallization and Evaporation

It is widely admitted that salt crystallization can be a major factor in the degradation of building materials, e.g. Scherer (2004) and references therein. An obvious step in order to assess the impact of crystallization is to first understand the ion migration during an evaporation process since the crystallization spots should correspond to the regions of high concentration where the ion concentration can reach the crystallization concentration, i.e. the concentration marking the onset of crystallization. As discussed in many works, e.g. Veran-Tissoires and Prat (2014) and references therein, the evaporation process

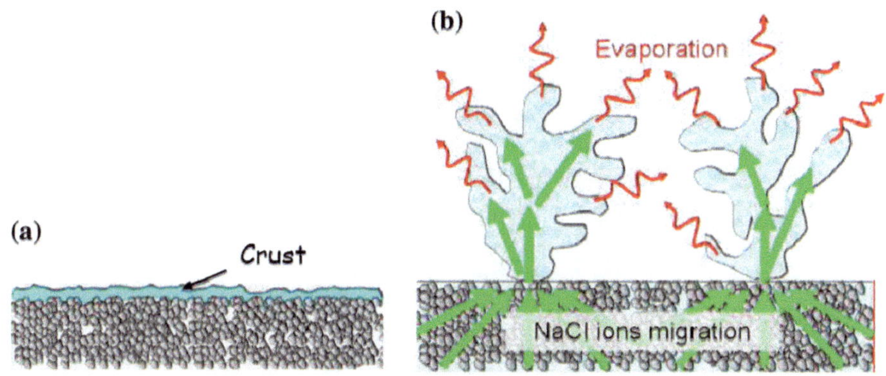


**Fig. 7.17** Sketch of two major types of salt efflorescence forming at the surface of a porous medium as a result of the combined action of evaporation, salt transport and crystallization: **a** crust, **b** cauliflower

induced a velocity field in the drying sample. Provided that the induced velocities are not too small (otherwise diffusion makes the ion concentration spatially uniform across the sample), the net result is the transport of the ions by the induced velocity field toward the evaporative surface. Depending of the drying period where crystallization occurs, salt can crystallize at the surface where it forms efflorescence or within the porous materials where it forms subflorescence. One question directly related to drying is the impact of crystallization on the drying process. The drying experiments with sodium chloride aqueous solutions reported in Eloukabi et al. (2013) led to distinguish two main situations regarding the impact of efflorescence. As sketched in Fig. 7.17, the efflorescence can either form a crust or cauliflower like well developed structures. The crust tends to severely limit the evaporation rate (drying becomes much slower than for pure water). By contrast, the cauliflower like structures, referred to as patchy efflorescence in Eloukabi et al. (2013), has practically no impact on the evaporation rate. As sketched in Fig. 7.17b, the cauliflower-like efflorescence is in fact porous and can transport the solution up to its external surface where water evaporates. As reported in Gupta et al. (2014), this change in the structure of efflorescence led to a paradoxical drying situation. In these experiments with again a sodium-chloride aqueous solution, it was found that drying was slower when the external relative humidity was set at 0 % than when the external relative humidity was set to 55 or 70 % in complete contrast with the situation with pure water. The explanation was that a salt crust formed for 0 % RH whereas the efflorescence had no impact on drying, thus of patchy—cauliflower type, for the two other relative humidities. The factor controlling the type of efflorescence are therefore the evaporation rate at the beginning of drying (a salt crust is expected to form for sufficiently high initial evaporation rates) and the mean pore size as reported in Eloukabi et al. (2013) (the probability of obtaining a crust increases with a decreasing mean pore size).

**Fig. 7.18** Image of a cauliflower salt efflorescence structure obtained by X-ray microtomography

As reported in Veran-Tissoires et al. (2012a, b), Hidri et al. (2013) and Veran-Tissoires and Prat (2014), the factors controlling the localisation of first crystals at the surface of porous medium subject to evaporation have been identified. They are the evaporation flux distribution (crystals are likely to form at the regions of surface where local evaporation fluxes are higher) and the disorder of porous medium microstructure (which is responsible for liquid velocity heterogeneities within the porous medium as well as its surface).

Despite these recent advances, there is still much to be done in this area, notably as regards the development of predictive models of drying in the presence of salt crystallization.

## 7.9 Digital Porous Media

Tools such as X-ray microtomographs tend to become increasing available for obtaining digital images of porous material microstructures, e.g. Bentz et al. (2000). As an example, Fig. 7.18 shows a cauliflower type efflorescence structure obtained using a X-ray microtomograph. The use of the corresponding thee dimensional digital images for characterizing transport properties of porous media or studying transport phenomena in porous media is referred to as digital porous media physics. Drying and related problems have also been studied thanks to this type of technique, e.g. Wang et al. (2012), Norouzi Rad et al. (2013). However, we believe that much remained to be done in this area since these techniques should enable one to assess the performances of drying models in a much more accurate way than from traditional macroscopic experiments.

## 7.10 Conclusions

As for other porous media, two main types of building materials can be distinguished in relation with the modelling of drying processes: capillary porous media and hygroscopic porous media. The continuum Darcy's scale models of drying still used nowadays were developed decades ago. It seems that these models have not been improved in recent years. Here we solely refer to the situations where the materials can be reasonably assumed as non-deformable. We are not considering the works devoted to the situations where drying occurs with cracks formation or mechanical deformations. It seems, however, that improvements of the continuum models are still needed for obtaining truly predictive models. Among the open questions, it seems necessary to develop a better modelling of mass transfer at the surface of a porous medium at the transition CRP/FRP and during the FRP in conjunction with the continuum approach.

By contrast, the last decade has seen quite a lot of works dealing with the modelling of drying directly at the pore scale or the pore network scale outside the Darcy's scale continuum framework. These approaches have significantly improved our understanding of drying at the considered scales. Confirming the insights gained from comparison between experiments on etched networks and pore network models obtained in the nineties, e.g. Laurindo and Prat (1998), the last decade has seen the modelling of liquid film in capillary porous media emerging has a major issue. Although the presently available pore network models of drying with liquid films led to good agreements with experiments on etched networks or with thin packing of beads confined between two plates, the development of an improved model seems necessary to reconcile the pore network simulations and experiments for fully three dimensional systems (as opposed to the quasi-two-dimensional experimental systems just mentioned).

The last decade has also seen significant advances in the understanding and modelling of drying in the presence of dissolved salts and crystallization. This is worth mentioning in relation with building materials since salt weathering is a major issue as regards the conservation of buildings and monuments.

## References

Assouline S, Narkis K, Gherabli R, Lefort P, Prat M (2014) Analysis of the impact of surface layer properties on evaporation from porous systems using column experiments and modified definition of characteristic length. Water Resources Research, WR014489

Bear J, Bachmat Y (1990) Introduction to modeling of transport phenomena in porous media. Springer, Berlin

Benet JC, Jouanna P (1983) Non equilibre thermodynamique dans les milieux poreux non satures avec changement de phase. Int J Heat Mass Transf 26:1585–1595

Bentz D, Quenard D, Baroghel-Bouny V, Garboczi E, Jennings H (1995) Modelling drying shrinkage of cement paste and mortar Part 1. Structural models from nanometres to millimetres. Mater Struct 28:450–458

Bentz DP, Garboczi EJ, Quenard DA (1998) Modelling drying shrinkage in reconstructed porous materials: application to porous Vycor glass. Modelling Simul Mater Sci Eng 6:211–236

Bentz DP, Quenard DA, Kunzel HM, Baruchel J, Peyrin F, Martys NS et al.(2000) Microstructure and transport properties of porous building materials. II: three-dimensional X-ray tomographic studies. Mater Struct 33:147–153

Blunt MJ, Jackson MD, Piri M, Valvatne PH (2002) Detailed physics. Predictive capabilities and macroscopic consequences for pore-network models of multiphase flow. Adv Water Resour 25:1069–1089

Carmeliet J, Descamps F, Houvenaghel G (1999) A multiscale network model for simulating moisture transfer properties of porous media. Transp Porous Media 35:67–88

Chauvet F, Duru P, Geoffroy S, Prat M (2009) Three periods of drying of a single square capillary tube. Phys Rev Lett 103:124502

Chen P, Pei DCT (1989) A mathematical model of drying processes. Int J Heat Mass Transf 32:297–310

Coussot P (2000) Scaling approach of the convective drying of a porous medium. Eur Phys J B 15:557–566

Defraeye T, Blocken B, Derome D, Nicolai B, Carmeliet J (2012) Convective heat and mass transfer modelling at air-porous material interfaces: overview of existing methods and relevance. Chem Eng Sci 74:49–58

Eijkel JCT, Dan B, Reemeijer HW, Hermes DC, Bomer JG, van den Berg A (2005) Strongly accelerated and humidity-independent drying of nanochannels induced by sharp corners. Phys Rev Lett 95:256107

Eloukabi H, Sghaier N, Ben Nasrallah S, Prat M (2013) Experimental study of the effect of sodium chloride on drying of porous media: The crusty—patchy efflorescence transition. Int J Heat Mass Transf 56:80–93

Gupta S, Huinink H, Prat M, Pel L, Kopinga K (2014) Paradoxical drying due to salt crystallization. Chem Eng, Sci

Hall C, Hoff WD (2002) In water transport in brick, stone and concrete. CRC Press, Boca Raton

Hidri F, Sghaier N, Eloukabi H, Prat M, Nasrallah SB (2013) Porous medium coffee ring effect and other factors affecting the first crystallisation time of sodium chloride at the surface of a drying porous medium. Phys Fluids 25:127101–127123

Kim J-H, Ochoa JA, Whitaker S (1987) Diffusion in anisotropic porous media. Transp Porous Media 2:327–356

Kohout M, Grof Z, Štěpánek F (2006) Pore-scale modelling and tomographic visualisation of drying in granular media. J Colloid Interface Sci 299:342–351

Kowalski SJ (2007) Continuous thermohydromechanical model using the theory of mixtures. In: Modern drying technology, Wiley-VCH Verlag GmbH & Co. KGaA, pp 125–154

Laurindo JB, Prat M (1998) Numerical and experimental network study of evaporation in capillary porous media. Drying rates. Chem Eng Sci 53:2257–2269

Le Bray Y, Prat M (1999) Three-dimensional pore network simulation of drying in capillary porous media. Int J Heat Mass Transf 42:4207–4224

Lehmann P, Assouline S, Or D (2008) Characteristic lengths affecting evaporative drying of porous media. Phys Rev E 77:056309

Lykov AV (1966) Heat and mass transfer in capillary-porous bodies, 1st English ed. Pergamon, Oxford

Mainguy M, Coussy O, Baroghel-Bouny V (2001) Role of air pressure in drying of weakly permeable materials. J Eng Mech 127:582–592

Masoero E, Del Gado E, Pellenq RJM, Ulm FJ, Yip S (2012) Nanostructure and nanomechanics of cement: polydisperse colloidal packing. J Phys Rev Lett 109:155503

Metzger T, Tsotsas E, Prat M (2007) Pore network models: a powerful tool to study drying at the pore level and understand the influence of structure on drying kinetics. In: Mujumdar A, Tsotsas ET (eds) Modern drying technology. Volume 1: computational tools at different scales, Wiley

Millington R (1959) Gas diffusion in porous media. Science 130:100–102

Norouzi Rad M, Shokri N, Sahimi M (2013) Pore-scale dynamics of salt precipitation in drying porous media. Phys Rev E 88:032404

Perré P, Rémond R, and Turner IW (2007) Comprehensive drying models based on volume averaging: background, application and perspective, chapter 1. In: Modern drying technology. Volume 1: computational tools at different scales, Wiley, pp 1–55

Perré P (2011) A review of modern computational and experimental tools relevant to the field of drying. Drying Technol 29:1529–1541

Philip J, DeVries D (1957) Moisture movement in porous materials under temperature gradients. Trans Am Geophys Union 38:222–232

Plumb OA (2000) Transport phenomena in porous media: modeling the drying process. In: Vafai K (ed) Handbook of porous media, Marcel Dekker Inc, New York, pp 755–785

Prat M (1993) Percolation model of drying under isothermal conditions in porous media. Int J Multiph Flow 19:691–704

Prat M, Bouleux F (1999) Drying of capillary porous media with stabilized front in two-dimensions. Phys Rev E 60:5647–5656

Prat M (2002) Recent advances in pore-scale models for drying of porous media. Chem Eng J 86:153–164

Prat M (2011) Pore network models of drying, contact angle and films flows. Chem Eng Technol 34:1029–1038

Scherer GW (2004) Stress from crystallization of salt. Cem Concr Res 34:1613–1624

Shahraeeni E, Lehmann P, Or D (2012) Coupling of evaporative fluxes from drying porous surfaces with air boundary layer: characteristics of evaporation from discrete pores. Water Resour Res 48:WR011857

Suzuki M, Maeda S (1968) On the mechanism of drying of granulars beds. J Chem Eng Jpn 1:26–31

Thiery M, Baroghel-Bouny V, Bourneton N, Villain G, Stéfani C (2007) Modélisation du séchage des bétons, analyse des différents modes de transfert hydrique. Revue européenne de Génie Civil 11:541–577

Veran-Tissoires S, Marcoux M, Prat M (2012a) Salt crystallisation at the surface of a heterogeneous porous medium. EPL (Europhys Lett) 98:34005

Veran-Tissoires S, Marcoux M, Prat M (2012b) Discrete salt crystallization at the surface of a porous medium. Phys Rev Lett 108:054502

Veran-Tissoires S, Geoffroy S, Marcoux M, Prat M (2013) Chapter 8: evaporation and wicking. In: Masoodi R, Pillai KM (eds) Wicking in porous materials: traditional and modern modeling approaches. CRC Press, Taylor & Francis

Veran-Tissoires S, Prat M (2014) Evaporation of a sodium chloride solution from a saturated porous medium with efflorescence formation. J Fluid Mechanics (in press)

Wang Y, Kharaghani A, Metzger T, Tsotsas E (2012) Pore network drying model for particle aggregates: assessment by X-ray microtomography. Drying Technol 30:1800–1809

Wang Y, Kharaghani A, Tsotsas E (2013) Pore scale simulation and visualization of liquid films for drying particle packings. In Proceedings of eurodrying, Paris

Whitaker S, James PH, Thomas FI (1977) Simultaneous heat, mass, and momentum transfer in porous media: a theory of drying. In: Advances in heat transfer, vol 13. Elsevier, pp 119–203

Whitaker S (1998) Coupled transport in multiphase systems: a theory of drying. Adv Heat Transf 31:1–103

Xu K, J-fo Daian, Quenard D (1997a) Multiscale structures to describe porous media part i: theoretical background and invasion by fluids. Transp Porous Media 26:51–73

Xu K, Daian J-F, Quenard D (1997b) Multiscale structures to describe porous media part ii: transport properties and application to test materials. Transp Porous Media 26:319–338

Yiotis AG, Salin D, Tajer ES, Yortsos YC (2012) Drying in porous media with gravity-stabilized fronts: experimental results. Phys Rev E 86:026310

# Chapter 8
# Cost Effective and Energy Efficient Control of Indoor Humidity in Buildings with Hygroscopic Building Materials and Desiccants in the HVAC System

**M. T. Fauchoux, C. J. Simonson, D. A. Torvi, R. M. Eldeeb and T. Ojanen**

**Abstract** Maintaining moderate indoor relative humidity levels in buildings is important for ensuring occupant comfort, as well as durability of building materials. This chapter presents methods for moderating indoor humidity levels using hygroscopic materials, in the form of: (1) hygroscopic building materials, (2) a desiccant coated air-to-air energy wheel and (3) a new concept ceiling panel with a vapour permeable surface. Simulation studies in an apartment and two office buildings, in different North American cities, are presented. The simulated indoor relative humidity and percent dissatisfied with perceived air quality (PAQ) are presented to show the benefits of using hygroscopic materials in buildings. Experimental results are presented for the novel ceiling panel, with results showing the sensible and latent effectiveness's, as well as the heat and mass flux rates obtained from the panel. The results of the simulations and experiments show that these devices and methods have the potential to improve thermal comfort and indoor air quality (IAQ), while reducing the energy consumption of the buildings.

M. T. Fauchoux (✉) · C. J. Simonson · D. A. Torvi · R. M. Eldeeb
Department of Mechanical Engineering, University of Saskatchewan, 57 Campus Drive, Saskatoon, SK S7N 5A9, Canada
e-mail: melanie.fauchoux@usask.ca

C. J. Simonson
e-mail: carey.simonson@usask.ca

D. A. Torvi
e-mail: david.torvi@usask.ca

R. M. Eldeeb
e-mail: rme490@mail.usask.ca

T. Ojanen
VTT Technical Research Centre of Finland, P.O. Box 1000, 02044 VTT, Espoo, Finland
e-mail: tuomo.ojanen@vtt.fi

J. M. P. Q. Delgado (ed.), *Drying and Wetting of Building Materials and Components*, Building Pathology and Rehabilitation 4, DOI: 10.1007/978-3-319-04531-3_8,

**Keywords** Indoor relative humidity • Hygroscopic materials • Energy efficiency • Thermal comfort

## 8.1 Introduction

Several researchers have found strong links between IAQ and the comfort, health and productivity of occupants (Wargocki et al. 2000a, b; Seppänen et al. 1999; Kosonen and Tan 2004). IAQ is a measure of the contaminants in the space air and can be improved by increasing the ventilation airflow rate to a space. Even if the IAQ of a space is acceptable, however, the occupants may perceive the air quality to be poor if the temperature and humidity of the space air are unacceptable. This is referred to as the perceived air quality (PAQ). Research has shown that high indoor humidity levels can significantly decrease the PAQ of a space (Fang et al. 1998a, b). In addition to affecting the PAQ of a space, the indoor relative humidity can significantly affect thermal comfort (Toftum et al. 1998a, b) and the health (Bornehag et al. 2001) of occupants.

Aside from the occupant aspect, humidity control in buildings is also important for structural purposes. Excess water vapour in the space air will condense onto, or absorb into, the surfaces in a space. Increased moisture in building materials can reduce the durability and lifetime of the materials and lead to mould growth (Trechsel and Bomberg 2009; Simonson et al. 2005; Hutcheon 1971; Chen et al. 2004; Moon and Yoon 2010). Poor indoor air conditions (high temperature and relative humidity) have also been found to increase the emission rates from building materials, in particular paint and varnish (Haghighat and de Bellis 1998). It is important to moderate indoor relative humidity levels in order to avoid these problems.

There are several methods currently employed for moderating indoor relative humidity levels, such as mechanical cooling equipment and desiccant dryers in humid climates and humidifiers or outdoor ventilation in dry climates, to humidify or dehumidify a space (Harriman et al. 1999; Pavlovas 2004). These systems typically consume large amounts of energy and require large initial costs for purchasing and installing the equipment. In addition, if the outdoor air is already too dry or too humid, it can make the indoor relative humidity levels worse. An alternative method is the use of hygroscopic materials to moderate the indoor relative humidity levels of a space. This chapter will look at three applications of hygroscopic materials in: (1) the building envelope, (2) the HVAC system (a desiccant coated air-to-air energy wheel) and (3) a new concept ceiling panel (referred to as a heat and moisture transfer panel or HAMP). This work shows that hygroscopic materials can be used to moderate indoor relative humidity, as well as improve PAQ while consuming less energy than typical systems.

## 8.2 Hygroscopic Building Materials

Hygroscopic building materials are materials that have the ability to absorb and release moisture, such as wood and gypsum. In a building, hygroscopic materials may be present in the walls, floor or ceiling of a space, as well as in the furniture. The potential for hygroscopic building materials to improve the indoor humidity and PAQ of a space is investigated using the LATENTITE computer model. The model combines the heat, air and moisture balance of the indoor air with the hygrothermal performance of the building envelope. The model has previously been presented and validated with field and laboratory experiments (Simonson et al. 2001, 2002, 2004a, b; Salonvaara 1998; Salonvaara and Karagiozis 1994; Simonson 2000).

To test the potential for hygroscopic building materials to improve the indoor relative humidity and PAQ, a bedroom in a wooden apartment building is simulated for three Canadian cities: Vancouver, B.C. (mild climate), Saskatoon, SK (cold and dry climate) and Toronto, ON (warm and humid climate). The bedroom is occupied by two adults, who produce 60 $g_w$/h of moisture and 90 W of sensible heat, for 9 h each night. The bedroom has a floor area of 12 $m^2$, with an internal surface area of 60 $m^2$ and a ventilation rate of 0.5 air changes per hour (ach). The wood frame structures are insulated with 150 mm of cellulose and have no plastic vapour retarder behind the interior board made of wooden panel. The walls, ceiling and floor are of the same lightweight construction; however, the floor is covered with 28 mm of wood coated with an impermeable coating. All of the building materials are permeable and hygroscopic. The property data for the construction is given in Simonson et al. (2001). Two cases are examined to investigate the difference between hygroscopic and non-hygroscopic materials. In the hygroscopic case, a vapour permeable paint [$5 \times 10^{-9}$ kg/(s $m^2$ Pa)] is applied to the inside surface of the structure. In the non-hygroscopic case, a vapour impermeable paint [$5 \times 10^{-12}$ kg/(s $m^2$ Pa)] is used. In the non-hygroscopic case, moisture transfer between the indoor air and the structure is reduced to essentially zero.

### 8.2.1 *Indoor Relative Humidity*

The simulated yearly variation in the indoor relative humidity, in Toronto, is shown in Fig. 8.1. In the non-hygroscopic case, the indoor relative humidity drops as low as 10 % RH in the winter, and increases to as high as 90 % RH in the summer. Similar results are obtained for Saskatoon (10–65 % RH) and Vancouver (20–75 % RH). There is a large amount of scatter in the indoor relative humidity throughout the year, in the non-hygroscopic case. The high indoor relative humidity levels mean that condensation may form on the windows and in the construction joints in the bedroom. In the hygroscopic case, the fluctuations in the indoor relative humidity

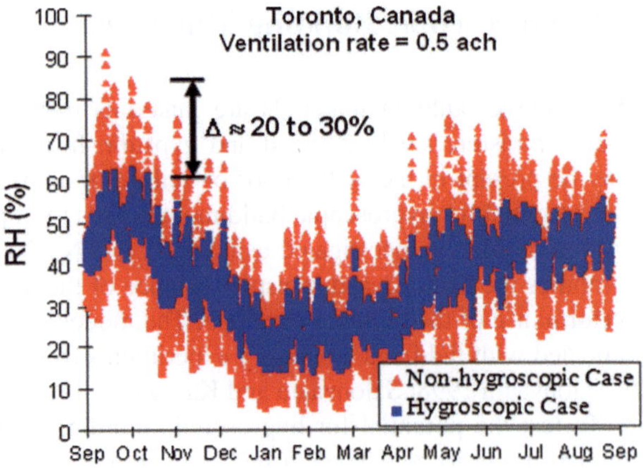


**Fig. 8.1** Effect of hygroscopic building materials on the indoor relative humidity in the bedroom of an apartment in Toronto, ON

are much smaller, with the indoor relative humidity reduced by about 15 % RH on average, in Toronto. The peak relative humidity reduction is 30 % RH in Toronto.

### *8.2.2 Perceived Air Quality*

Studies by Fang et al. (1998a, b) have shown that temperature and relative humidity have a strong impact on the PAQ of a space. They have developed equations to calculate the percentage of people that are dissatisfied (PD) with the PAQ ($PD_{PAQ}$) based on the acceptability of the air. The correlation is dependent on the amount of contaminants in the air. For clean air, the correlation is:

$$PD_{PAQ} = \frac{\exp(-0.18 - 5.28Acc)}{1 + \exp(-0.18 - 5.28Acc)} 100 \quad (8.1)$$

where Acc is the acceptability of the air, calculated using the equation for facial exposure to clean air (Fang et al. 1998b). The equation for acceptability is:

$$Acc = -0.033H + 1.662 \quad (8.2)$$

where H is the enthalpy of the air (kJ/kg).

The percent dissatisfied with PAQ was calculated for the case with hygroscopic materials and the case without hygroscopic materials in the bedroom, for each city. The change in the percent dissatisfied with PAQ ($\Delta PD_{PAQ}$) was then calculated as the difference between these two values. The change in $\Delta PD_{PAQ}$ is shown in Fig. 8.2, as a frequency graph, for all three cities. It can be seen that the addition of hygroscopic building materials to the bedroom improves the percent dissatisfied during almost all nights of the year ($\Delta PD_{PAQ} > 0$). In some cases, the improvement in $\Delta PD_{PAQ}$ is as high as 20 or 30 % when hygroscopic materials are used in the bedroom. On average, the peak daily value of $PD_{PAQ}$ is reduced by 4–7 %. These results show that it is possible to improve the PAQ of occupied spaces, using hygroscopic building materials in the building envelope.

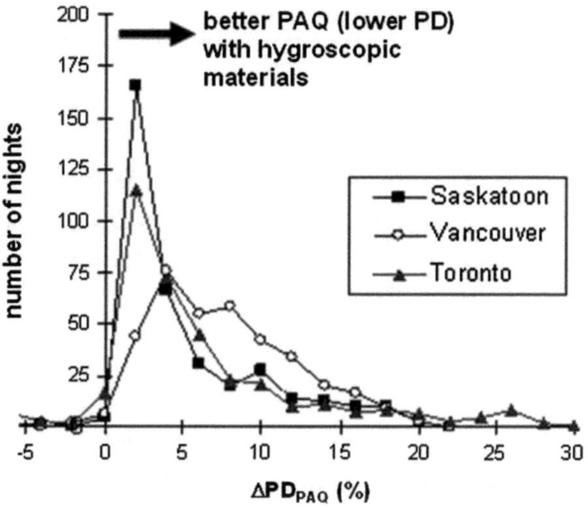


Fig. 8.2 Improvement in the percent dissatisfied with perceived air quality in the bedroom, in each city

### *8.2.3 Energy and Economics*

Hygroscopic materials, such as wood and gypsum, are materials that are needed for structural or aesthetic purposes, so the initial cost of installing these materials is absorbed into the cost of constructing the building. A reduction in the $PD_{PAQ}$ of a space has been linked to an increase in productivity and occupant health (Wargocki et al. 2000a). An increase in productivity and occupant health can lead to labour cost savings, as well as higher revenues of a company. Based on these results, hygroscopic building materials are a very cost effective means of improving indoor relative humidity levels.

## 8.3 Desiccant Coated Air-to-Air Energy Wheel

It has been demonstrated that hygroscopic building materials can help to improve the indoor relative humidity levels in a building. Hygroscopic materials can also be used in the HVAC system of a building, specifically, in the form of a desiccant coated air-to-air energy wheel. Energy wheels are commonly used in buildings, as a means of reducing energy consumption. They also have the ability however, to moderate in the indoor relative humidity of a space, by adding or removing moisture from the air before it is supplied to the space. To show the effects of an energy wheel on the indoor relative humidity, the TRNSYS computer program (Solar Energy Laboratory 2005) is used to model an office building in four North American cities: Saskatoon, SK (cold and dry climate), Vancouver, B.C. (mild climate), Tampa, FL (hot and humid climate) and Phoenix, AZ (hot and dry climate).

The 15-storey office building has a curtain wall construction, with windows covering 40 % of the wall area. The U-value of the walls is 0.21 W/($m^2$ K) and the

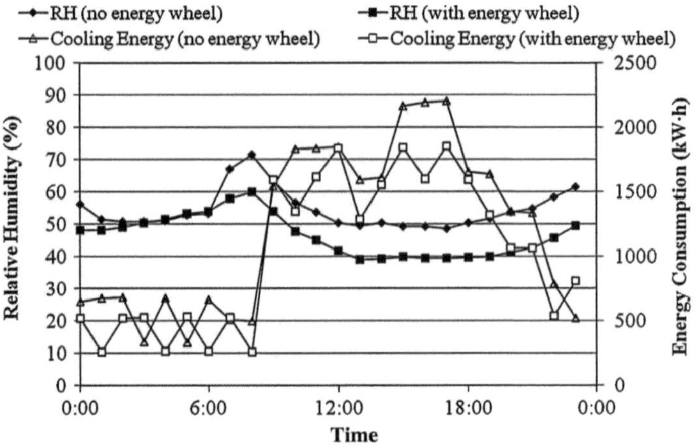


**Fig. 8.3** Simulated indoor relative humidity and cooling energy consumption with and without a desiccant coated energy wheel in Tampa, FL, on August 22

U-value of the windows is 2.7 W/($m^2$ K). The building has 1,600 occupants and a floor area of 23,000 $m^2$. The building has an infiltration rate of 0.3 ACH and an outdoor ventilation rate of 16,000 L/s (ASHRAE Standard 62.1 2010). The HVAC system is a constant volume all-air system. The supply air consists of 20 % fresh outdoor air under normal conditions, however when the outdoor conditions are favourable, this percentage is increased to minimize energy consumption. The supply air is heated with a natural gas boiler (79 % efficiency) or cooled with an electrical driven cooling unit (COP of 2.7) (ASHRAE Standard 90.1 2010). Complete details of the office building and HVAC system are provided by Fauchoux (2006) and Fauchoux et al. (2007). The office building is simulated with and without a desiccant coated energy wheel (70 % sensible and latent effectiveness's) to determine the effect of hygroscopic materials on the indoor relative humidity.

### *8.3.1 Indoor Relative Humidity*

The hourly indoor relative humidity in the office building is shown in Fig. 8.3, for Tampa, FL on August 22. The results show that the energy wheel is able to reduce the peak indoor relative humidity in the building, at almost every hour of the day. The indoor relative humidity is reduced by 10–15 % RH during most of the occupied period. The energy wheel removes some of the moisture from the outdoor air before it is delivered to the cooling coil and space, reducing the amount of moisture that the supply air adds to the space, thus reducing the indoor relative humidity.

Figure 8.4 shows the difference in the indoor relative humidity with and without an energy wheel, as a frequency distribution, in the four cities. The difference in the indoor relative humidity is calculated using:

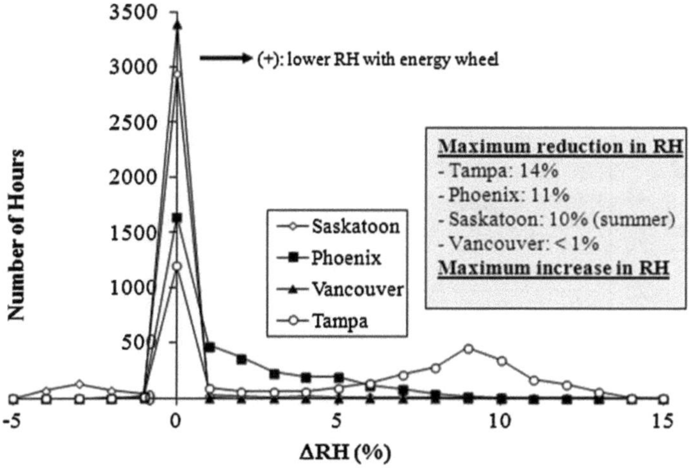


**Fig. 8.4** Comparison of the indoor relative humidity calculated with and without a desiccant coated energy wheel during occupied hours

$$\Delta RH = RH_{\text{without energy wheel}} - RH_{\text{with energy wheel}}. \tag{8.3}$$

ΔRH is plotted over 1 % RH intervals, so a value of zero is actually in the range of −0.5–0.5 % RH. The largest impact is seen in Tampa, where the indoor relative humidity is reduced by >2 % RH during 61 % of the occupied hours when the energy wheel is used. The reduction in indoor relative humidity is >10 % RH during 34 % of the occupied hours, with the maximum reduction being 14 % RH. In Phoenix, the indoor relative humidity is reduced by >2 % RH during 37 % of the occupied hours, with a maximum reduction of 11 % RH.

The indoor relative humidity is considerably lower in Saskatoon, than in Tampa or Phoenix, due to the dry climate. In Saskatoon, it is favourable to increase the indoor relative humidity during the winter months, in addition to decreasing the indoor relative humidity during the summer months. The maximum increase in the winter is 5 % RH and the maximum decrease in the summer is 10 % RH. In Vancouver (mild climate), the energy wheel has very little effect on the indoor relative humidity.

### *8.3.2 Perceived Air Quality*

The hourly $PD_{PAQ}$ is shown in Fig. 8.5 for Tampa on August 22. As with the indoor relative humidity, the energy wheel is able to reduce the peak value of $PD_{PAQ}$ in the building significantly and reduce the $PD_{PAQ}$ at almost every hour throughout the day.

The difference in the $PD_{PAQ}$ with and without an energy wheel ($\Delta PD_{PAQ}$) is shown in Fig. 8.6, for all four cities. $\Delta PD_{PAQ}$ is calculated in the same manner

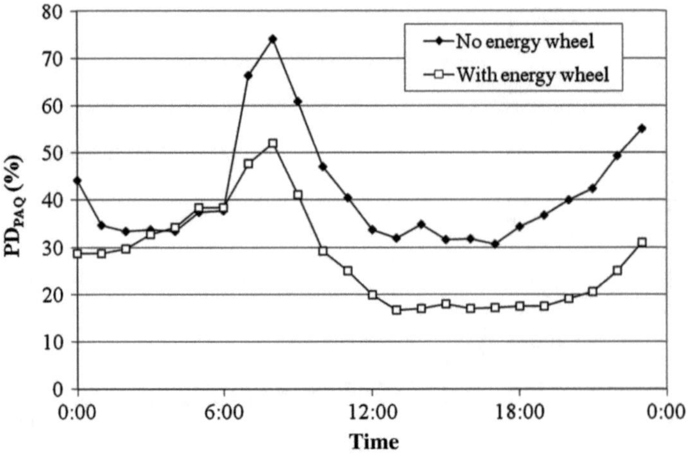


**Fig. 8.5** Simulated percent dissatisfied with perceived air quality with and without a desiccant coated energy wheel in Tampa, FL, on August 22

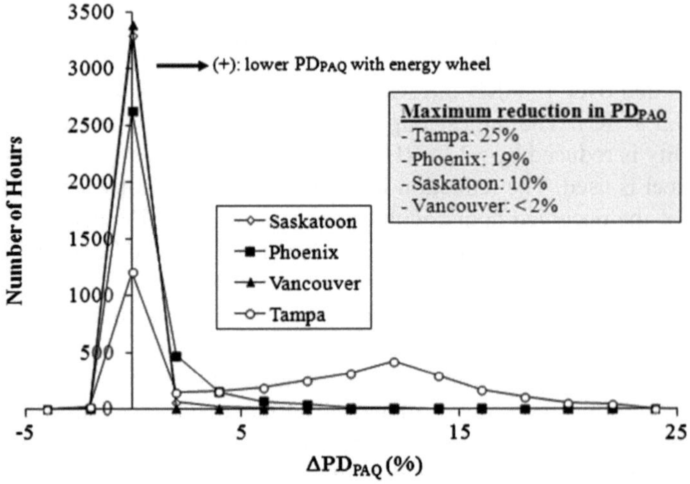


**Fig. 8.6** Comparison of the $PD_{PAQ}$ calculated with and without a desiccant coated energy wheel during occupied hours

as $\Delta RH$ (Eq. 8.3). The largest impact is seen in Tampa, where the addition of an energy wheel to the HVAC system is able to reduce the $PD_{PAQ}$ by up to 25 %. The maximum reduction in the $PD_{PAQ}$ is 19 % in Phoenix and 10 % in Saskatoon. Again, there is no significant impact on the $PD_{PAQ}$ in Vancouver when an energy wheel is used.

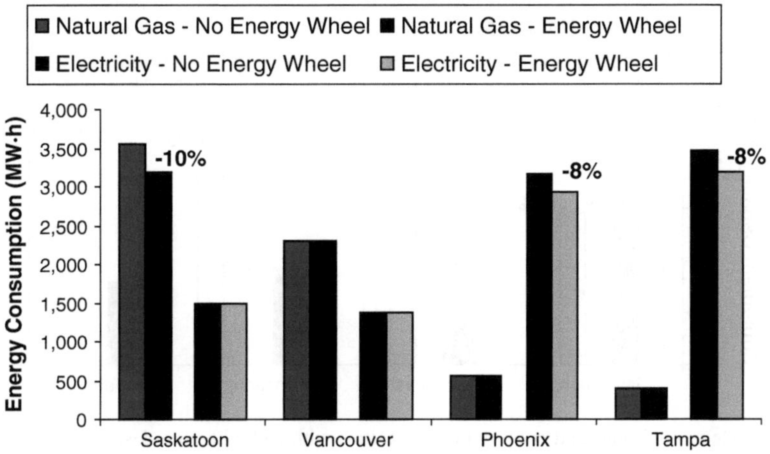


**Fig. 8.7** Annual energy consumption of the HVAC system with and without a desiccant coated energy wheel

### *8.3.3 Energy Consumption and Equipment Capacity*

Energy wheels are commonly used in HVAC systems to reduce the energy consumption required for conditioning the supply air for buildings. As such, the total energy consumption and required equipment capacities for the two cases studied are presented to show that acceptable indoor relative humidity levels can be maintained in buildings without increasing the energy consumption of the building.

The total amount of natural gas (heating) and electricity (cooling and fans) consumed annually by the HVAC system, with and without an energy wheel, is shown in Fig. 8.7, for each city. Since Saskatoon is a cold climate, the majority of energy consumed is used for heating. With the addition of an energy wheel to the HVAC system, the natural gas consumption decreases by 10 % as compared to the case with no energy wheel. In Phoenix and Tampa (hot climates), the majority of energy consumed is used for cooling. With the addition of an energy wheel to the HVAC system, the electrical energy consumed is reduced by 8 % in both cities.

The required capacity of the boiler and cooling unit is shown in Fig. 8.8, for the case with and without the energy wheel, in each city. In Saskatoon, the addition of a hygroscopic energy wheel to the HVAC system reduces the boiler capacity by 26 % and the cooling unit capacity by 4 %. In Phoenix and Tampa, the addition of an energy wheel reduces the cooling unit capacity by 18 and 17 %, respectively. There are no significant changes in the capacity of the heating or cooling units in Vancouver. A life cycle cost analysis, comparing the energy savings to the initial capital costs of the equipment, shows payback periods of less than one year in Phoenix and Tampa. In Saskatoon, the payback period is negative due to the

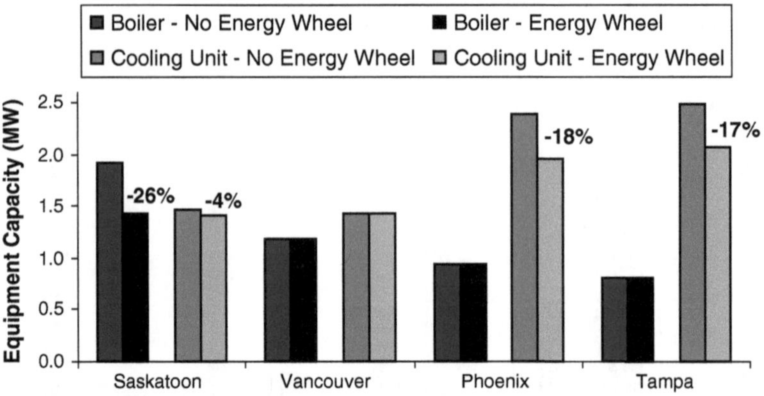


**Fig. 8.8** Required boiler and cooling unit capacities with and without a desiccant coated energy wheel

large reductions in the capacities of the heating and cooling units. This means that the payback in Saskatoon is immediate and any benefits (energy savings, improved indoor relative humidity and PAQ) result from essentially no investment. Since there are virtually no energy savings in Vancouver, the payback period is infinitely large.

## 8.4 A Heat and Moisture Transfer Panel

Radiant heating and cooling panel systems can be used in buildings to control the temperature of a space. These systems consist of a temperature-controlled surface on the floor, walls or ceiling of the space. Heat transfer occurs through thermal radiation between the panel and the various surfaces in the space, as well as through convection between the panel and the air in the space. Several researchers (Leskinen 2003; Imanari et al. 1999; Niu et al. 2002; Vangtook and Chirarattananon 2006; ASHRAE 2012) have documented the benefits of using radiant heating and cooling panels. Radiant panel systems provide improved indoor air quality, better thermal comfort for occupants and consume less energy annually than typical all-air systems. In addition, the amount of supply air required for the space is reduced to the amount required for ventilation standards. For new buildings, this means reduced duct sizes, smaller plenums and reduced equipment capacities for pre-conditioning the supply air. One of the limiting factors of existing radiant panels is that they can only transfer sensible heat with the space, and not moisture. In humid climates, when the panels are used for cooling, there is a risk of condensation forming on the panels. In dry climates, there is no way for the panels to add moisture to the space, to improve thermal comfort.

A solution to this problem is the development of a heat and moisture transfer panel (HAMP) which has the ability to transfer both heat and moisture with a space.

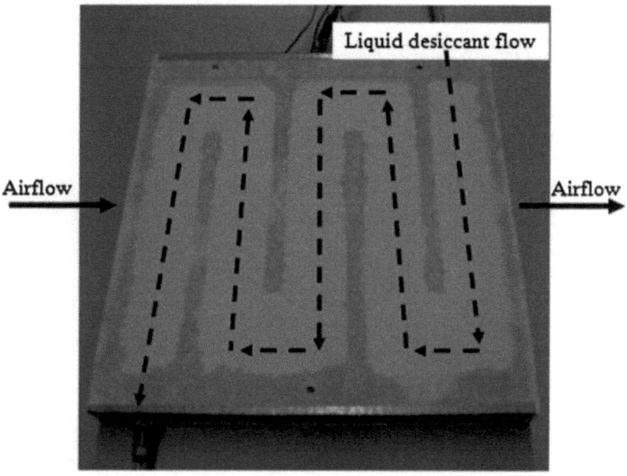


**Fig. 8.9** Photograph of the prototype HAMP with directions of the air and liquid desiccant flows shown

As with a radiant panel, a HAMP is built into the surface of a space, however, with a HAMP, both the temperature and humidity ratio of the surface are controlled. The surface of a HAMP is made of a semi-permeable membrane (such as Propore™, Tyvek®) which is permeable to water vapour, to allow moisture transfer, but impermeable to liquid water, to contain the working fluid within the HAMP. The working fluid in a HAMP is a liquid desiccant, allowing a HAMP to transfer heat and moisture simultaneously, in order to heat or cool and humidify or dehumidify a building space, as needed. A HAMP is another example of a hygroscopic material being used in the surface of a space to moderate the indoor relative humidity.

### *8.4.1 Prototype HAMP*

To determine the feasibility of a HAMP, a prototype was created and tested in a small wind-tunnel facility. The prototype HAMP, shown in Fig. 8.9, is made of an acrylic tray with dimensions of 27.9 cm × 25.4 cm × 3.2 cm high. The acrylic tray is divided into five channels, to ensure the liquid desiccant flows evenly from inlet to outlet. The direction of the air and liquid desiccant flows are shown in Fig. 8.9. The effective surface area of the prototype for heat and moisture transfer is 0.043 $m^2$. The height of the liquid desiccant flow in the channels is 2.5 cm.

The surface of the prototype HAMP is Propore™, which is a micro-semi-permeable polypropylene membrane, affixed to a non-woven polypropylene fabric for support. Propore™ has a low vapour diffusion resistance (high moisture transfer through the membrane), a high airflow resistance (low airflow passing through the membrane) and a high liquid penetration pressure (the pressure at which liquid will leak through the membrane) (Larson 2006). The liquid desiccant used in the prototype HAMP is an aqueous salt solution. An aqueous salt solution has a lower surface vapour pressure than pure water, providing a lower surface relative

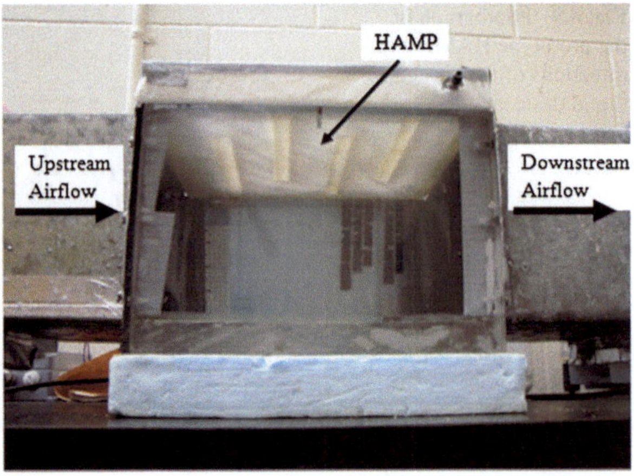


**Fig. 8.10** Photograph of the prototype HAMP in the ceiling of the test section

humidity. This enables the liquid desiccant to remove moisture from the space air, in addition to humidifying the air, as needed. The prototype HAMP uses an aqueous Lithium Chloride solution, which provides a surface relative humidity of 12 % RH (1.7 $g_w/kg_{air}$) at saturated conditions, in a 20 °C solution (ASTM Standard E104 2007). The concentration of the salt solution in the prototype is varied between 24 and 60 % to create different moisture gradients between the HAMP and the airflow. The temperature and density of the liquid desiccant are measured during each test. The humidity ratio of the liquid desiccant is calculated from these values using correlations (Cisternas and Lam 1991).

The prototype HAMP is placed into the test section of a small wind tunnel facility as shown in Fig. 8.10. The test section is a rectangular duct, with dimensions of 15.2 cm high by 22.9 cm wide, which gives a hydraulic diameter of 18.3 cm. The air entering the test section is hydrodynamically fully developed, laminar flow. The bulk temperature and relative humidity of the air, upstream and downstream of the test section, are measured during each test. The HAMP has been tested as a ceiling panel, however the design of the test facility is such that the HAMP could be tested as a floor or wall panel as well. Full details of the prototype HAMP and the test facility are presented in Fauchoux (2012), Fauchoux et al. (2008, 2009).

The range of test conditions used to test the performance of the prototype HAMP under different operating conditions are given in Table 8.1. The HAMP was tested for heating, cooling, humidifying and dehumidifying of the airflow. The temperature of the upstream air was kept at approximately room temperature for all tests. The temperature of the liquid desiccant inside the HAMP varied between 6 and 35 °C to test cases of both cooling and heating of the air. For humidifying cases, the relative humidity of the upstream air was set to a low value and the surface relative humidity of the liquid desiccant was set to a high value to create a large potential for moisture transfer. For dehumidifying cases, the relative humidity of the upstream air was set to a high value and the surface relative humidity of the liquid desiccant to a low value.

**Table 8.1** Summary of test conditions used to test the performance of the prototype HAMP

| Parameter | Lower limit | Upper limit |
|---|---|---|
| $T_{UPSTREAM}$ (°C) | 20 | 25 |
| $RH_{UPSTREAM}$ (% RH) | 3 | 87 |
| $W_{UPSTREAM}$ ($g_w/kg_{air}$) | 0.5 | 16 |
| $T_{HAMP}$ (°C) | 6 | 35 |
| $\rho_{HAMP}$ (kg/m$^3$) | 1,141 | 1,398 |
| $C_{HAMP}$ (%) | 24 | 60 |
| $RH_{HAMP}$ (% RH) | 15 | 61 |
| $W_{HAMP}$ ($g_w/kg_{air}$) | 2 | 28 |

**Table 8.2** Sensible and latent effectiveness's of the prototype HAMP under different operating conditions

| Operating conditions | $\varepsilon_{sensible}$ (%) | $\varepsilon_{latent}$ (%) |
|---|---|---|
| Cooling, dehumidifying | $12 \pm 5$ | $48 \pm 3$ |
| Cooling, humidifying | $16 \pm 5$ | $46 \pm 5$ |
| Heating, dehumidifying | $6 \pm 5$ | $23 \pm 7$ |
| Heating, humidifying | $4 \pm 5$ | $27 \pm 2$ |
| Dehumidifying | – | $30 \pm 6$ |
| Humidifying | – | $26 \pm 3$ |

### *8.4.2 Effectiveness of the Prototype HAMP*

The ability of the prototype HAMP to transfer heat is quantified by a sensible effectiveness, calculated using:

$$\varepsilon_{sensible} = \frac{T_{DOWNSTREAM} - T_{UPSTREAM}}{T_{HAMP} - T_{UPSTREAM}} \tag{8.4}$$

where T is the temperature (°C) of the upstream air, downstream air or the liquid desiccant inside the HAMP, based on the subscript. The ability of the prototype HAMP to transfer moisture is quantified by a latent effectiveness, calculated using:

$$\varepsilon_{latent} = \frac{W_{DOWNSTREAM} - W_{UPSTREAM}}{W_{HAMP} - W_{UPSTREAM}} \tag{8.5}$$

where W is the humidity ratio ($g_w/kg_{air}$).

The sensible and latent effectiveness's of the prototype HAMP during six tests, representing different operating conditions, are presented in Table 8.2. The tests considered are a case of cooling and dehumidifying, cooling and humidifying, heating and dehumidifying, heating and humidifying, dehumidifying (with no heating or cooling) and humidifying (with no heating or cooling). The experimental uncertainty of each value is also presented for each test. The results show that the performance of the prototype HAMP varies significantly for each set of operating conditions. It was determined that there are several factors that influence the performance of the HAMP, including the presence of buoyancy forces during some tests, as well as the coupling of the heat and moisture transfer between the HAMP and the air (Fauchoux et al. 2010a, 2011).

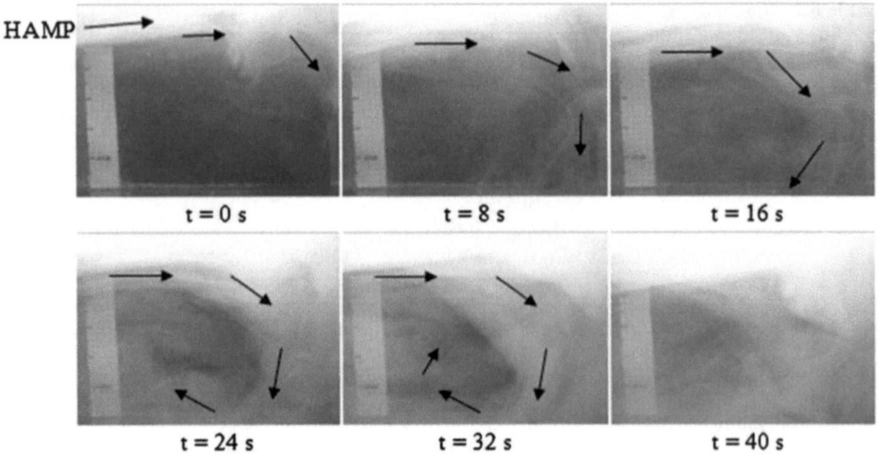


**Fig. 8.11** Visualization of unstable laminar airflow through the test section during a case of cooling and dehumidifying (Fauchoux et al. 2010b)

When the HAMP is used for cooling and/or dehumidifying of the air, buoyancy forces are present in the test section, due to the density gradient created across the height of the test section (high density at the top, low density at the bottom). The buoyancy forces cause convection roll cells to form in the test section, which increases mixing of the air and therefore, increases the amount of heat and moisture transferred between the HAMP and the air. This results in higher sensible and latent effectiveness's when the HAMP is used for cooling and/or dehumidifying. Figure 8.11 shows the presence of a convection roll cell in the test section during a case of cooling and dehumidifying.

When the HAMP is used for heating and/or humidifying of the air, there are no buoyancy forces present in the test section, because the density of the air at the top of the test section is lower than the density of the air at the bottom of the test section. In these cases, the airflow remains stable and therefore, the amount of heat and moisture transferred between the HAMP and the air is lower than in the cooling and dehumidifying cases. This results in lower sensible and latent effectiveness's during cases of heating and/or humidifying.

### *8.4.3 Total Heat and Mass Fluxes*

The performance of a radiant ceiling panel is typically presented as the total heat flux between the panel and the space air at different air and panel temperatures. As the HAMP is able to transfer both heat and moisture with the space air, the total mass flux under different conditions is also of interest. The total heat flux between the HAMP and the air in the test section is calculated using:

**Fig. 8.12** Total heat flux for the prototype HAMP for both heating and cooling conditions

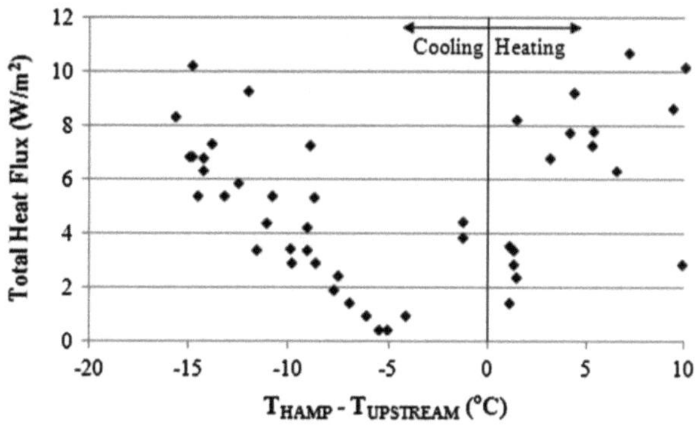


$$q'' = U\Delta T_{lm} \tag{8.6}$$

where U is the overall heat transfer coefficient [W/(m$^2$ K)] and $\Delta T_{lm}$ is the log mean temperature difference between the airflow and the liquid desiccant (°C). The log mean temperature difference is calculated using:

$$\Delta T_{lm} = \frac{\Delta T_{in} - \Delta T_{out}}{\ln\left(\Delta T_{in}/\Delta T_{out}\right)} \tag{8.7}$$

where $\Delta T_{in}$ is the temperature difference between the HAMP and the air at the entrance of the test section and $\Delta T_{out}$ is the temperature difference between the HAMP and the air at the outlet of the test section (K).

The total mass flux between the HAMP and the air is calculated using:

$$\dot{m}'' = U_m \Delta W_{lm} \tag{8.8}$$

where $U_m$ is the overall mass transfer coefficient [$kg_{air}$/(m$^2$ s)] and $\Delta W_{lm}$ is the log mean humidity ratio difference between the airflow and the liquid desiccant ($g_w/kg_{air}$), calculated from Eq. (8.7) by replacing the temperature with the humidity ratio.

The total heat flux between the HAMP and the air in the test section is shown in Fig. 8.12 for both heating and cooling cases. The total heat flux is plotted against the difference between the temperature of the liquid desiccant inside the HAMP and the airflow. The total heat flux increases as the difference in the temperatures increases. Compared to a typical radiant ceiling panel, the total heat flux between the HAMP and the air is low. This is due to the low velocities used in the experiments, which results in low convective heat transfer rates. An increase in the convective heat transfer coefficient would increase the overall heat flux of the HAMP.

The total mass flux between the prototype HAMP and the air in the test section is shown in Fig. 8.13 for both dehumidifying and humidifying conditions. The total mass flux is plotted against the difference in the humidity ratio of the liquid desiccant inside the HAMP and the humidity ratio of the airflow. The total mass flux between the HAMP and the airflow increases as the difference in the humidity ratios increases. A person seated, doing light work, as would be the case

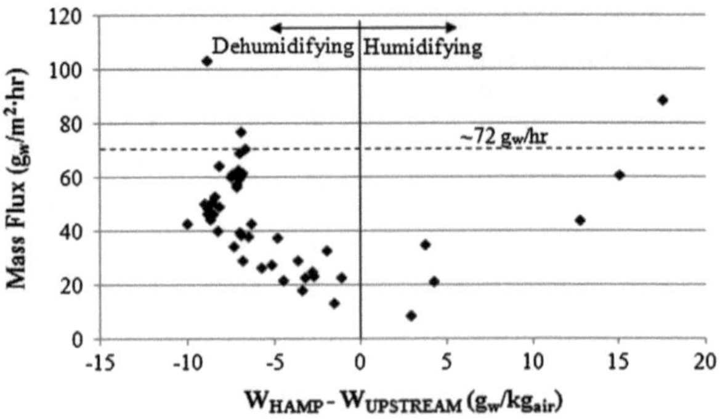


**Fig. 8.13** Total mass flux for the prototype HAMP for both dehumidifying and humidifying conditions

in a typical office building, produces 45 W of latent heat (ASHRAE 2009). This amount of latent heat generation equates to a production of 72 $g_W$/h per person. Depending on the difference in the humidity ratios, a HAMP with a surface area of approximately 1–2 $m^2$ would be required to remove the moisture produced by one person. This is an appropriate amount for a space with a low occupant density, such as an office building.

### *8.4.4 Indoor Relative Humidity*

The TRNSYS building simulation program is used to investigate the performance of a HAMP in a 1-storey office building in Saskatoon, SK (cold and dry climate) and Chicago, IL (cold and humid climate) (Eldeeb et al. 2012; Eldeeb 2012). The building has a floor area of 730 $m^2$ with wall, roof, floor and fenestration constructions based on ASHRAE Standard 90.1 (2010) for the climate zones that represent each city. The occupancy density in the building is 5 people/100 $m^2$, the maximum lighting intensity is 10.8 W/$m^2$ and the maximum equipment intensity is 10.7 W/$m^2$. The infiltration rate is calculated at each time step, with an average value of 0.28 ach.

To determine the feasibility of a HAMP in an office building, three cases are simulated, as described in Table 8.3. Case A represents a conventional all-air system, which removes the space and ventilation sensible loads, as well as the ventilation latent load. In case B, radiant ceiling panels are used to remove the space sensible load, instead of the variable air volume (VAV) system. Mechanical ventilation is used to remove the sensible and latent ventilation loads. Finally, case C uses a HAMP to remove the latent space load, in addition to the radiant ceiling panels that remove the sensible space load and mechanical ventilation to remove the sensible and latent ventilation loads.

The radiant ceiling panels are modelled as an active layer in the building with a user defined fluid specific heat and the heat transfer coefficient of the panel. The heat transfer coefficient is calculated at every time step based on the panel temperature and the space air temperature, using correlations from ASHRAE (2012). The

**Table 8.3** Description of the three cases simulated to determine the performance of a HAMP

| Case | Space load | | Ventilation load | |
|---|---|---|---|---|
| | Sensible | Latent | Sensible | Latent |
| A | VAV | None | VAV | VAV |
| B | RCP | None | MV | MV |
| C | RCP | HAMP | MV | MV |

*VAV* variable air volume; *RCP* radiant ceiling panel; *MV* mechanical ventilation; *HAMP* heat and moisture transfer panel

**Table 8.4** Liquid desiccant set points for each operating condition

| Operating conditions | $T_{HAMP}$ (°C) | $C_{HAMP}$ (%) | $W_{HAMP}$ ($g_w/kg_{air}$) |
|---|---|---|---|
| Heating and humidification | 40 | 38 | 12.1 |
| Heating and dehumidification | 35 | 50 | 4.6 |
| Cooling and humidification | 16 | 6 | 10.9 |
| Cooling and dehumidification | 16 | 32 | 4.2 |

area of the radiant ceiling panels is 60 % of the total ceiling area. The set point temperature in the space is 24 °C for cooling and 22 °C for heating.

The moisture transfer between the HAMP and the space is modelled as a moisture gain to the space (positive for moisture added to the air, negative for moisture removed from the air). This gain is calculated using a Matlab code with input information from the building space, using the equation:

$$\dot{m} = U_m A \rho_{dryair}(W_{HAMP} - W_{air}) \tag{8.9}$$

where Um is the overall convective mass transfer coefficient (m/s), A is the surface area of the HAMP ($m^2$), ρ is the density of dry air ($kg_{air}/m^3$) and W is the humidity ratio ($g_w/kg_{air}$). The area of the HAMP is 10 % of the total ceiling area. In order to control the indoor temperature and relative humidity of the space, both the temperature and humidity ratio of the liquid desiccant inside the HAMP must be set. Table 8.4 lists the set point conditions for the HAMP for each operating condition.

A frequency graph of the relative humidity in the office building, during occupied hours, is shown in Fig. 8.14 for all three cases, in Chicago. In case A, the indoor relative humidity ranges from 3 to 65 % RH. This shows that the all-air system is effective at dehumidifying the air, but not humidifying the air. There are a significant number of hours where the indoor relative humidity is less than 30 % RH. In case B, the indoor relative humidity ranges from 3 to 78 % RH. There is less dehumidification of the ventilation air in this case, resulting in higher indoor humidity levels. Again, there is no humidification of the air, so the indoor relative humidity is low for a large number of hours. Case C, which uses a HAMP to control the space latent load, in addition to mechanical ventilation to handle the ventilation latent load, maintains the indoor relative humidity between 26 and 62 % RH for all occupied hours. There are peaks at 35 and 60 % RH, which correspond

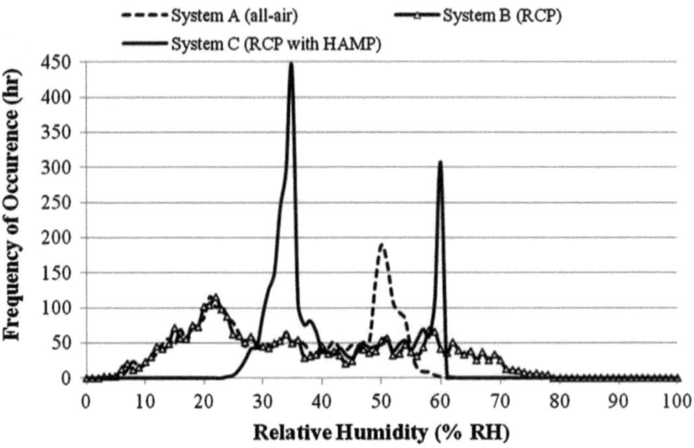


**Fig. 8.14** Frequency of the indoor relative humidity inside a one-storey office building in Chicago, IL

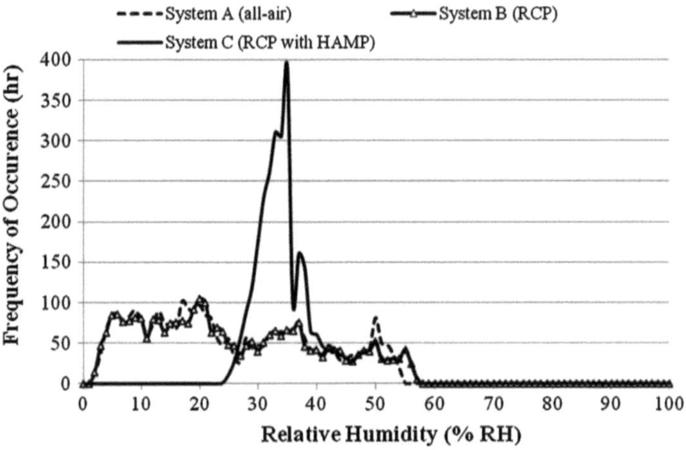


**Fig. 8.15** Frequency of the indoor relative humidity inside a one-storey office building in Saskatoon, SK

to the set points humidity values for the HAMP. This shows that a HAMP has the potential to improve the indoor relative humidity levels as compared to two traditional HVAC systems.

A frequency graph of the relative humidity inside the office building, during occupied hours, is shown in Fig. 8.15 for the three cases, in Saskatoon. In case A, the indoor relative humidity ranges from 1 to 55 % RH. Since Saskatoon is a very dry climate, only a small amount of dehumidification is needed. There are a large number of hours however where the indoor relative humidity is very low, suggesting humidification is necessary. In case B, the indoor relative humidity is in the range of 1–61 % RH, very similar to case A. In case C, when the HAMP is used, the indoor relative humidity is maintained within the range of 24–57 % RH for all occupied hours. Again, this shows the benefits of using a HAMP compared to the other two systems.

### *8.4.5 Energy Consumption*

The experimental results presented, and the simulated results from TRNSYS show that a HAMP has the potential to add/remove moisture from the air in a space, as needed, in order to maintain the indoor relative humidity within acceptable limits. At this point, the energy consumption required by a HAMP has not been determined. Some energy will be required for pumping the liquid desiccant through the HAMP, which will be similar to the energy required for pumping water through a radiant ceiling panel. As moisture is added to or removed from the liquid desiccant inside the HAMP, the concentration of the liquid desiccant will change, which will change its ability to absorb or release more moisture. In addition to the energy required to pump the liquid desiccant through the HAMP, some energy will be required to regenerate the liquid desiccant to the desired salt concentration. If low-grade energy sources, such as solar energy are used for regeneration, the energy consumption of a HAMP is expected to be very low. Research into the amount of energy required for regeneration is still ongoing.

## 8.5 Conclusions

Three different forms of hygroscopic materials have been presented to show the potential for hygroscopic materials to control indoor humidity levels and improve PAQ of spaces. Simulations are performed with and without hygroscopic finishing materials in a bedroom, with and without hygroscopic materials in the HVAC system of a building and with and without a moisture transfer panel in the ceiling of a space. In all three cases, the hygroscopic materials were able to improve the indoor relative humidity of the spaces in hot and humid climates, as well as cold and dry climates. In a mild climate, such as Vancouver, the hygroscopic building materials resulted in improved indoor relative humidity levels; however, the energy wheel had little effect on the indoor humidity levels. The greatest effects were seen in hot and humid climates.

An energy consumption and life-cycle costs analysis on the energy wheel showed that acceptable indoor humidity levels could be provided without requiring a considerable amount of energy and cost. The use of hygroscopic materials in the building envelope can provide acceptable indoor conditions with virtually no investment or energy consumption, as most of the materials are commonly used in building structures already. Finally, the energy consumption of a HAMP is expected to be similar to radiant ceiling panels, which have been shown to consume less energy than all-air systems. In summary, hygroscopic building materials are a cost effective and energy efficient means to provide adequate indoor humidity levels to achieve good comfort and prevent deterioration of buildings.

## References

ASHRAE (2009) Fundamentals. American Society of Heating, Refrigerating and Air-conditioning Engineers Inc., Atlanta

ASHRAE (2012) HVAC systems and equipment. American Society of Heating, Refrigerating and Air-conditioning Engineers, Inc., Atlanta

ASHRAE Standard 62.1 (2010) Ventilation for acceptable indoor air quality. American Society of Heating, Refrigerating and Air-conditioning Engineers, Inc., Atlanta

ASHRAE Standard 90.1 (2010) Energy standard for buildings except low-rise residential buildings. American Society of Heating, Refrigerating and Air-conditioning Engineers, Inc., Atlanta

ASTM Standard E104 (2007) Standard practice for maintaining constant relative humidity by means of aqueous solutions. American Society for Testing and Materials International, West Conshohocken, PA

Bornehag CG, Blomquist G, Gyntelberg F, Jarvholm B, Malmberg P, Nordvall L, Nielsen A, Pershagen G, Sundell J (2001) Dampness in buildings and health. Indoor Air 11:72–86

Chen H, Deng S, Bruner Jr H, Garcia J (2004) Roots of mold problems and humidity control measures in institutional buildings with pre-existing mold condition. In: Proceedings of the fourteenth symposium on improving building systems in hot and humid climates, Richardson, Texas, May 2004, paper ESL-HH-04-05-07

Cisternas LA, Lam EJ (1991) Analytic correlation for the vapour pressure of aqueous and non-aqueous solutions of single and mixed electrolytes: part II—application and extension. Fluid Phase Equilib 62:11–27

Eldeeb RM (2012) Simulating a heat and moisture transfer panel (HAMP) for maintaining space humidity. M.Sc. thesis, University of Saskatchewan, Saskatoon

Eldeeb RM, Fauchoux MT, Simonson CJ (2012) TRNSYS modeling of a novel ceiling panel designed to maintain space humidity in an office building. eSim 2012 Building simulation conference, Halifax

Fang L, Clausen G, Fanger PO (1998a) Impact of temperature and humidity on perception of indoor air quality during immediate and longer whole-body exposures. Indoor Air 8:276–284

Fang L, Clausen G, Fanger PO (1998b) Impact of temperature and humidity on perception of indoor air quality. Indoor Air 8:80–90

Fauchoux M (2006) The effect of energy recovery on indoor climate, air quality and energy consumption using computer simulations. M.Sc. thesis, University of Saskatchewan, Saskatoon. http://library2.usask.ca/theses/available/etd-06222006-162448/

Fauchoux M (2012) Design and performance testing of a novel ceiling panel for simultaneous heat and moisture transfer to moderate indoor temperature and relative humidity. Ph.D. thesis, University of Saskatchewan, Saskatoon. http://ecommons.usask.ca/handle/10388/ETD-2012-10-747

Fauchoux M, Simonson CJ, Torvi DA (2007) The effect of energy recovery on perceived air quality, energy consumption and the economics of an office building. ASHRAE Trans 113(2):437–449

Fauchoux MT, Simonson CJ, Torvi DA (2008) Investigation of a novel ceiling panel for heat and moisture control in buildings. In: Rode C (ed) 8th Nordic symposium on buildings physics, Copenhagen, Denmark, pp 1269–1276

Fauchoux MT, Simonson CJ, Torvi DA (2009) Tests of a novel ceiling panel for maintaining space relative humidity by moisture transfer from an aqueous salt solution. J ASTM Int 6(4)

Fauchoux M, Bansal M, Talukdar P, Simonson CJ, Torvi D (2010a) Testing and modeling of a novel ceiling panel for maintaining space relative humidity by moisture transfer. Int J Heat Mass Transf 53:3961–3968

Fauchoux MT, Simonson CJ, Torvi DA (2010b) Flow visualization of airflow through a rectangular duct with combined heat and mass transfer. In: Proceedings of the 14th international heat transfer conference, Washington, DC

Fauchoux MT, Simonson C, Torvi D, Talukdar P (2011) CFD modelling with buoyancy effects for a heat and moisture transfer ceiling panel. In: Proceedings of the ASME/JSME 8th thermal engineering joint conference, Honolulu, Hawaii

Haghighat F, de Bellis L (1998) Material emission rates: literature review, and the impact of indoor air temperature and relative humidity. Build Environ 33(5):261–277

Harriman LG, Witte MJ, Czachorski M, Kosar DR (1999) Evaluating active desiccant systems for ventilating commercial buildings. ASHRAE J 41(10):28–37

Hutcheon NB (1971) Buildings for the Canadian climate. National Research Council of Canada, Division of Building Research. Technical report 336

Imanari T, Omori T, Bogaki K (1999) Thermal comfort and energy consumption of the radiant ceiling panel system—comparison with the conventional all-air system. Energy Build 30:167–175

Kosonen R, Tan F (2004) The effect of perceived indoor air quality on productivity loss. Energy Build 36:981–986

Larson M (2006) The performance of membranes in a newly proposed run-around heat and moisture exchanger. M.Sc. thesis, University of Saskatchewan, Saskatoon. http://library.usask.ca/theses/available/etd-12192006-094159/

Leskinen M (2003) Low Exergy guidebook: heating and cooling with focus on increased energy efficiency and improved comfort. Technical Research Centre of Finland, Espoo

Moon HJ, Yoon YR (2010) Investigation of physical characteristics of houses and occupants' behavioural factors for mould infestation in residential buildings. Indoor Built Environ 19:57–64

Niu JL, Zhang LZ, Zuo HG (2002) Energy savings potential of chilled-ceiling combined with desiccant cooling in hot and humid climates. Energy Build 34:487–495

Pavlovas V (2004) Demand controlled ventilation: a case study for existing Swedish multifamily buildings. Energy Build 36:1029–1034

Salonvaara MH (1998) Prediction of hydrothermal performance of building envelope parte couple with indoor climate. ASHRAE Trans 104(2):908–918

Salonvaara M, Karagiozis A (1994) Moisture transport in building envelopes using an approximate factorization solution method. In: Proceedings of the second annual conference of the CFD Society of Canada, Toronto, pp 317–326

Seppänen O, Fisk W, Mendell M (1999) Association of ventilation rates and $CO_2$ concentrations with health and other responses in commercial and institutional buildings. Indoor Air 9:226–252

Simonson CJ (2000) Moisture, thermal and ventilation performance of Tapanila ecological house. VTT Research Notes, No. 2069, Technical Research Centre of Finland, Espoo. http://www.vtt.fi/inf/pdf/tiedotteet/2000/T2069.pdf

Simonson CJ, Salonvaara M, Ojanen T (2001) Improving indoor climate and comfort with wooden structures. VTT Publications, No. 431, Technical Research Centre of Finland, Espoo. http://www.vtt.fi/inf/pdf/publications/2001/P431.pdf

Simonson CJ, Salonvaara M, Ojanen T (2002) The effect of structures on indoor humidity—possibility to improve comfort and perceived air quality. Indoor Air 12:243–251

Simonson CJ, Salonvaara M, Ojanen T (2004a) Moisture performance of an airtight vapor-permeable building envelope in a cold climate. J Therm Envelope Build Sci 28(2):161–185

Simonson CJ, Salonvaara M, Ojanen T (2004b) Moderating indoor conditions with hygroscopic building materials and outdoor ventilation. ASHRAE Trans 110(2):804–819

Simonson CJ, Ojanen T, Salonvaara M (2005) Moisture performance of an airtight vapor-permeable building envelope in a cold climate. J Therm Envelope Build Sci 28(3):205–226

Solar Energy Laboratory (2005) A TRaNsient systems simulation program. University of Wisconsin, Madison. http://sel.me.wisc.edu/trnsys

Toftum J, Jorgensen AS, Fanger PO (1998a) Upper limits for indoor air humidity to avoid uncomfortably humid skin. Energy Build 28:1–13

Toftum J, Jorgensen AS, Fanger PO (1998b) Upper limits of air humidity for preventing warm respiratory discomfort. Energy Build 28:15–23

Trechsel HR, Bomberg MT (eds) (2009) Moisture control in buildings: the key factor in mold prevention, 2nd edn. ASTM International, Pennsylvania

Vangtook P, Chirarattananon S (2006) An experimental investigation of application of radiant cooling in hot humid climate. Energy Build 38:273–285

Wargocki P, Wyon DP, Fanger PO (2000a) Productivity is affected by the air quality in offices. Proc Healthy Build 1:635–640

Wargocki P, Wyon DP, Sundell J, Clausen G, Fanger PO (2000b) The effects of outdoor air supply rate in an office and perceived air quality, sick building syndrome (SBS) symptoms and productivity. Indoor Air 10:222–236

# Chapter 9
# Risk of Condensation in Mechanically Attached Roof Systems in Cold U.S. Climate Zones

**Simon Pallin and Manfred Kehrer**

**Abstract** Flat roofs have generally a high potential of nightly overcooling and therefore an increased risk of condensation within the construction, particularly in cold climate zones, depending on their specific assembly. A white exterior surface, so-called "cool roof", applied to decrease cooling loads and to save energy, increases the condensation risk due to lower energy gains at day time. The nightly overcooling phenomena will be intensified and leads to much lower temperature at the exterior surface compared to a standard roof. Observations (Energy Design Update® 2006) show that the increased condensation risk may lead to moisture damage. There have been questions raised about the sustainability of using cool roof membranes in Northern US climate zones due to the potential of moisture accumulation below the membrane. Transient hygrothermal simulation using real climate data are state of the art today and can help to study different effects. Variations of several input parameter, such as short-wave absorptivity of the solar radiation, ventilation underneath the membrane with interior or exterior air, specific climate data, etc., show their influences on the moisture accumulation underneath the membrane of a typical mechanically attached roof systems for commercial buildings. Consequently, the most significant input parameter can be determined and used as additional criteria for a better design.

**Keywords** Condensation · Overcooling · Hygrothermal simulation · Long-wave radiation · WUFI · Moisture accumulation

S. Pallin (✉)
Chalmers University of Technology, Chalmers, Sweden
e-mail: simon.pallin@chalmers.se

M. Kehrer
Oak Ridge National Laboratory, Oak Ridge, TN, USA
e-mail: kehrerm@ornl.gov

J. M. P. Q. Delgado (ed.), *Drying and Wetting of Building Materials and Components*, Building Pathology and Rehabilitation 4, DOI: 10.1007/978-3-319-04531-3_9,

## 9.1 Background

A white roof, or *cool roof*, is constructed to decrease thermal loads from solar radiation, therefore saving energy by decreasing the cooling demands. Unfortunately, cool roofs with a mechanically attached membrane have shown a higher risk of intermediate condensation in the materials below the membrane in certain climates (Ennis and Kehrer 2011) and in comparison with similar constructions with a darker exterior surface (Bludau et al. 2009). As a consequence, questions have been raised regarding the sustainability and reliability of using cool roof membranes in northern U.S. climate zones.

A white roof surface reflects more of the incident solar radiation than a dark surface, making a distinguishable difference in the surface temperature of the roof. However, flat roofs facing a clear sky, with either a light or a dark surface, are constantly losing energy to the sky due to the exchange of infrared radiation. This phenomenon exists both during the night and the day. During the day, if the sun shines on the roof surface, the exchange of infrared radiation typically becomes insignificant. During nights and in cold climates, the temperatures of the roof surface and the sky can differ by as much as 20 °C (Hagentoft 2001), which could result in a roof surface temperature that is much colder than the ambient temperature. Further, a colder roof surface has increased energy loss and risk of condensation in the building materials below the membrane. In conclusion, both light- and dark-coated roof membranes are cooled by infrared radiation exchange during the night, though a darker membrane is heated more by solar radiation during the day, thus decreasing the risk of condensation.

The phenomenon of nighttime cooling from contact with the sky and the lack of solar gain during the day is not likely the exclusive issue affecting the risk of condensation in cool roofs with mechanically attached membranes. Roof systems with thermoplastic membranes are prone to be more affected by interior air intrusion into the roof construction, both from wind-induced pressure differences and from the flexibility and elasticity of the membrane (Molleti et al. 2011). Depending on the air permeability of the material underneath the membrane, wind forces increase the risk of fluttering (also referred to as billowing) of a flexible single ply thermoplastic membrane. Expectably, the wind-induced pressure differences create a convective air flow into the construction (i.e., air intrusion). If the conditions are right, moisture from the exchanging air may condensate on surfaces with a temperature below the dew-point.

The definite path of convective air flows through the building envelope is usually very difficult to determine, so simplified models (Künzel et al. 2011) help to estimate the additional moisture loads caused by air intrusion. The wind uplifting pressure in combination with wind gusts is an important factor for a fluttering roof. Unfortunately, the effect of fluctuating wind is difficult to estimate as this is a highly dynamic phenomenon and existing standards (ASTM 2011a) take into account only a steady-state approach (i.e., there are no guidelines or regulations on how to estimate the air intrusion rate). Obviously, more detailed knowledge on the

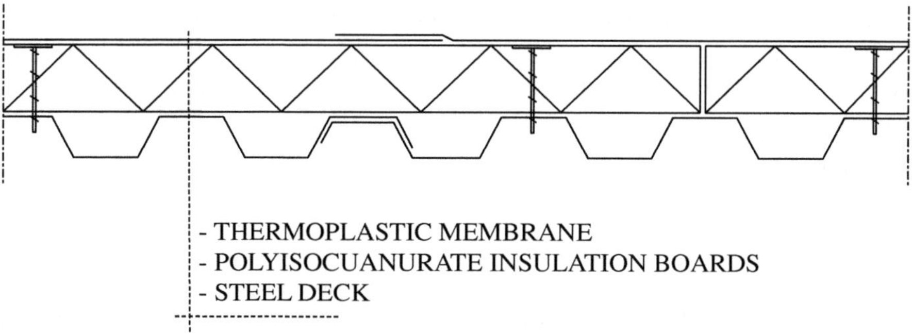


**Fig. 9.1** Model of the investigated mechanically attached roof system. The thermoplastic membrane has either a light or a dark surface that affects the hygrothermal performance of the roof

hygrothermal performance of mechanically attached cool roof systems is needed with regard to surface colors, roof airtightness, climate zones, and indoor moisture supply.

## 9.2 Introduction

The hygrothermal performance of the above-mentioned mechanically attached roof system has been investigated with numerical simulations. Measurements of the airtightness of the roof construction are necessary (see Sect. 9.4) to provide reliable input data for simulations. Studying these phenomena requires a tool capable of modeling heat and moisture transport in a transient simulation and with realistic boundary conditions. Long-wave (infrared) radiation must be considered at the exterior surface, otherwise nightly cooling cannot be taken into account in the simulations of a cool roof. On account of these prerequisites, the hygrothermal software WUFI is used for computations of coupled heat and moisture transport (Künzel 1995). WUFI, which has been validated repeatedly (Kehrer and Schmidt 2008), features a detailed radiation model, based on physical fundamentals, which calculates the temperatures of the exterior surfaces, thus determining the risk of condensation. In consideration of the expected air intrusion underneath the thermoplastic roof membrane, the air exchange model of WUFI is applied to estimate the heat and moisture exchange of indoor air.

Figure 9.1 shows the modeled assembly, consisting of a traditional metal deck, 3-inch polyisocuanurate insulation boards, and a thermoplastic membrane (representing a flexible single ply membrane). Due to a supposedly air leaky roof construction underneath the thermoplastic membrane, an uplift of the exterior membrane will lead to indoor air intrusion. As a consequence of the air intrusion, an air layer is created underneath the membrane.

The risk of condensation in the investigated roof assembly in Fig. 9.1 depends on several parameters. Their influences on hygrothermal performance are analyzed in a systematic parameter study in which the following input parameters are varied:

- Climate, including wind, solar loads and nighttime cooling
- Indoor moisture supply
- Air intrusion rate

Varying the parameters leads to 128 different combinations (i.e., scenarios) that are simulated in WUFI. The composition of the combinations and the different variations of input parameters are discussed in Sect. 9.4.

Further, good workmanship is assumed, resulting in a tightly sealed thermoplastic membrane. The mechanical resistance of the membrane is neglected; that is, any applied pressure on top of the membrane is immediately, and without resistance, equalized underneath the membrane.

## 9.3 Investigation and Measurements

Air intrusion into a roof construction depends on the wind loads acting on the roof surface and the air permeability of the construction below the thermoplastic membrane. Subsequently, air intrusion is an important parameter to estimate when analyzing the hygrothermal performance of a roof. This section presents an approach to quantify this parameter.

### *9.3.1 Wind Forces*

The airtightness of a roof construction, together with the existing pressure differences, determines the air leakage rate between the indoor and outdoor environment. In this study, the thermoplastic membrane is assumed to be sealed with satisfactory workmanship, thus minimizing exfiltration or infiltration air exchange; that is, air will not flow between the inner and outer surfaces of the roof. Instead, the exchange of air inside and between the roof materials is a consequence of indoor air intrusion (i.e., an exchange of indoor air in the materials of the building envelope). The air pressure differences influencing the intrusion rate are typically thermally driven (stack effect), with the ventilation system or wind loads acting on the building envelope (Hagentoft 2001). The wind generally creates an uplifting force on the thermoplastic membrane. Due to the flexibility and elasticity of the membrane, it may easily deform in favor of pressure differences, causing the membrane to flutter and balloon (Baskaran and Molleti 2010). This deformation, due to uplifting forces, is also referred to as billowing (Fig. 9.2).

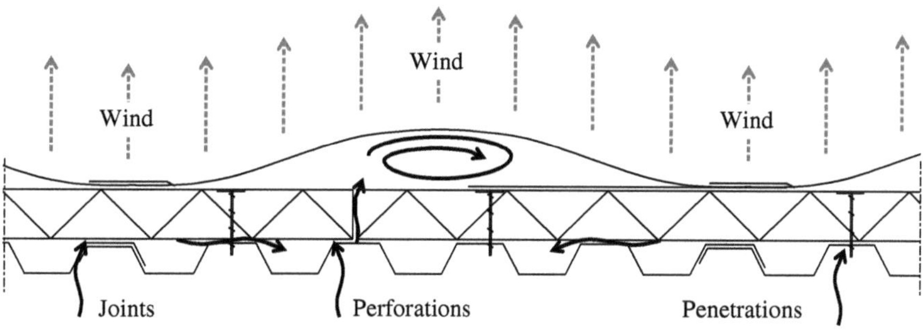


**Fig. 9.2** Wind forces inducing pressure differences on the outer roof membrane, causing the membrane to flutter and balloon. Depending on the condition and workmanship of the construction, air intrusion may arise in overlapping joints of the steel deck, penetrations, or perforations. Plausible locations of air intrusion are indicated with *solid arrows*. The uplifting forces, due to the wind, are indicated with *dashed arrows*

Two essential parameters must be specified to estimate the air intrusion rate in a roof assembly. One is wind speed fluctuation, which causes a fluctuating pressure difference between the outdoor and indoor surfaces of the roof, and the other is airtightness of the roof construction below the outer membrane.

A negative wind-induced air pressure is typically acting on the exterior roof surface and can be estimated by the use of ASCE Standard ASCE/SEI 7-10 (ASCE 2010) in which the wind pressure, Pw, for low-rise buildings is defined in Sects. 28.3.2 and 28.4.1. Following the calculation steps of the ASCE Standard and determining the suitable coefficients for the roof construction, exposure and surrounding topography result in an uplifting pressure, Pw, as follows:

$$P_w = -0.273 \cdot V^2 \tag{9.1}$$

where $V$ (m/s) is the wind speed parallel to the surface.

The wind speed is typically presented as an average speed for a defined period of time (e.g., 1 h). Wind speeds that are based on measured averages at shorter time intervals are referred to as gusts, Vg (Harper et al. 2010). Therefore, V can be expressed as

$$V = \frac{1}{m} \cdot \sum_{i=1}^{m} V_g^i, \tag{9.2}$$

in which $m$ is the number of gust samples within 1 h. In Fig. 9.3, the variations of $V$ and $V_g$ are presented for a period of 6 h, extracted from minute-based measurements of the wind speed in Holzkirchen, Germany from 2009 to 2010.

The difference in wind speed, ΔV, defines the pressure difference. A shorter time step between measured wind speeds is decisive for determining the fluttering effect of the roof membrane, and thus the air intrusion. Unfortunately, climate files usually consist of hourly averages of wind speed in which the differences between

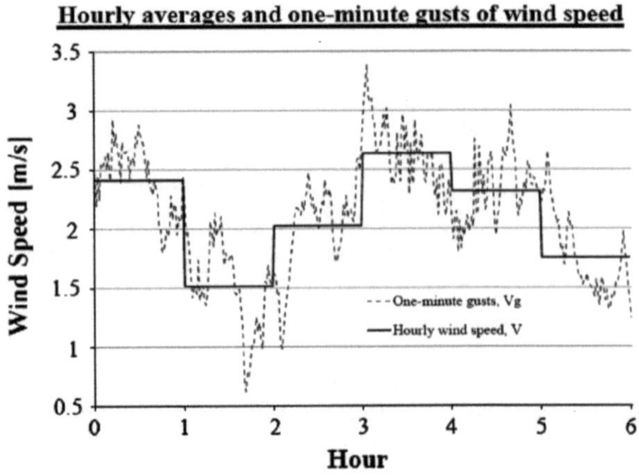


**Fig. 9.3** Variations of 1-minute wind gusts in Holzkirchen, Germany from 2009 to 2010 and the resulting hourly average wind speeds during the extracted measurements of 6 h

high and low wind speeds at shorter time periods are lost, as revealed in Fig. 9.3. Therefore, the hourly averages of data used in this study are adjusted to a normalized minute-based variation of the wind speed. This procedure estimates a ΔV at a given hourly wind speed, intended to estimate the uplifting pressure difference, Δpw, acting on the outer membrane surface. The minute-based gust measurements from Holzkirchen serve as the template for such adjustments, because of their known reliability and continuity. In conclusion, the hourly averages of wind speed given from the four U.S. climate zones are adjusted to instead represent a plausible variation of wind speed, ΔV, at each given hour.

The hourly average of ΔV is defined as

$$\Delta V = \frac{1}{m-1} \cdot \sum_{i=1}^{m} \left| V_g^{i+1} - V_g^i \right| \tag{9.3}$$

where $m$ is the number of gust measurements during 1 h. In this study, $V_g$ is expressed each minute; hence $m = 60$ and $i$ is the number of consecutive measurements.

The relative deviation in wind gusts typically decrease with increasing average wind speed (Davis and Newstein 1968), which also has been verified with the minute-based gust measurements from Holzkirchen, as seen in Fig. 9.4. This effect has also been implemented into the estimation of the wind induced pressure difference over the roof membrane.

Finally, the pressure and uplifting force acting on the membrane surface, $\Delta P_w$, is expressed using (ASCE 2010) and (ASTM 2011a).

$$\Delta P_w = -0,273 \cdot \left( \sum_{i=1}^{m} \left| V_g^{i+1} - V_g^i \right| \right)^2 \tag{9.4}$$

Due to the decrease in relative deviation of the wind gusts with increasing hourly average wind speed (as seen in Fig. 9.4), Eq. (9.4) provides different values of ΔPw at different hourly average wind speed.

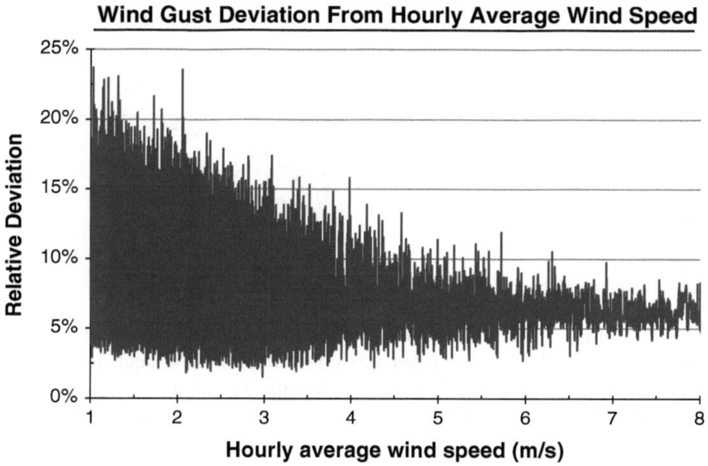


**Fig. 9.4** The relative deviation in maximum and minimum wind gusts typically decreases with increasing hourly average wind speed

## *9.3.2 Air Intrusion*

The fluttering effect of the roof sheeting causes an exchange of indoor air underneath the membrane. This air intrusion is defined as an air exchange rate per hour, $ACH$, and can be defined as

$$ACH = \frac{Q}{V} \tag{9.5}$$

where $Q$ is the air flow rate ($m^3$/h) and $V$ is the volume of air space that is ventilated ($m^3$), later assumed to be constant in the numerical model for practical reasons.

The variable $Q$, in this study, depends on the pressure difference between the membrane and the indoor environment and also on the airtightness of the roof assembly, as given by

$$Q = A \cdot C \cdot \Delta P^n, \tag{9.6}$$

where $A$ is the roof surface area ($m^2$); $C$ is the air leakage coefficient ($m^3$/s Pa); $\Delta P$ is the pressure difference (Pa), which in this study equals $\Delta P_w$; and finally $n$ is the pressure exponent (−). The $Q_{50}$-value (l/s or l/s $m^2$) refers to air flow rate at a pressure difference of 50 Pa, thus simplifying comparisons between different constructions and measurements.

Consequently, $C$ and $n$ are parameters related to the physical structure of the roof which are typically determined by measurements. Therefore, the roof construction defined in Sect. 9.2 is tested in accordance with the ASTM E2178-11 Standard for testing the air leakage rate (ASTM 2011b). The specimen represents the roof construction as defined in Fig. 9.1, except for the exclusion of the outer thermoplastic membrane. The reason for this approach is that the complete roof

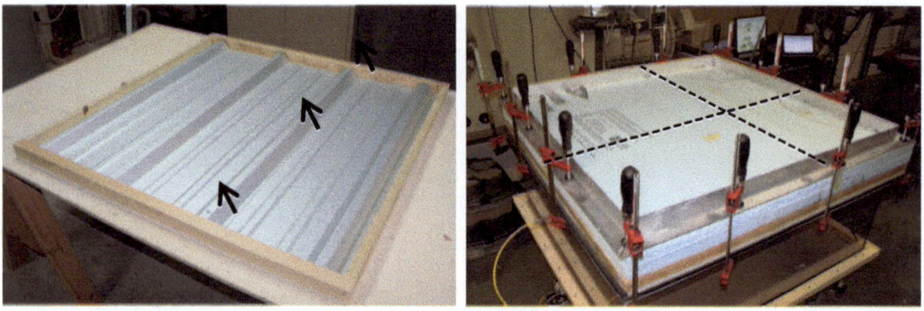

**Fig. 9.5** The airtightness of the materials below the thermoplastic membrane, which was tested by use of ASTM E2178-11. The specimen was constructed with a steel deck including a joint and, on top, two overlapping 1.5-inch insulation boards. The screws, ensuring a tight overlap, are indicated with *arrows*, and the joints between the overlapping insulation boards are indicated with *dashed lines* in the *right-hand* picture

**Table 9.1** Results from measuring the air leakage coefficient, C, and the pressure exponent, n

| Results from airtightness tests | $C$ (m$^3$/s Pa) | $n$ (−) | $Q_{50}$ (m$^3$/s) | $Q_{50}$ (l/s) |
|---|---|---|---|---|
| 1. Sealed joints and sealed screw penetrations | 3.31E−07 | 0.99 | 1.74E−05 | 0.02 |
| 2. Steel deck only | 6.45E−06 | 0.95 | 2.63E−04 | 0.26 |
| 3. Full assembly | 6.19E−06 | 0.96 | 2.69E−04 | 0.27 |
| 4. Full assembly, 2–4 perforations | 6.85E−05 | 0.54 | 5.61E−04 | 0.56 |
| 5. Full assembly, 8 perforations | 1.25E−04 | 0.72 | 2.09E−03 | 2.09 |

Five different assemblies were measured with various repetitions and are presented as average values of the measurements

assembly is assumed to be very airtight and that the air intrusion rate between the indoor environment and underneath the membrane is to be estimated.

Figure 9.5 displays the constructed roof specimen. The steel deck includes one overlapping joint and has been screwed tight at three positions along the overlapping ridge, as indicated with solid arrows in the picture below. Further, two layers of overlapping 1.5-inch insulation boards are mounted on top of the wood-framed steel deck.

The roof specimen was tested in five different assemblies.

1. Sealed joints and sealed screw penetrations
2. Steel deck only
3. Full assembly
4. Full assembly, two to four 3/16-inch steel deck perforations
5. Full assembly, eight 3/16-inch steel deck perforations

The results of the measurements are presented in Table 9.1 and Fig. 9.6, in which the different assemblies were measured with various repetitions. The purpose of the first assembly (sealed joints and sealed penetrations) was to ensure a satisfactory seal between the steel deck and the wooden framework, which is confirmed by the results in Table 9.1. Second, only the steel deck was tested, without the

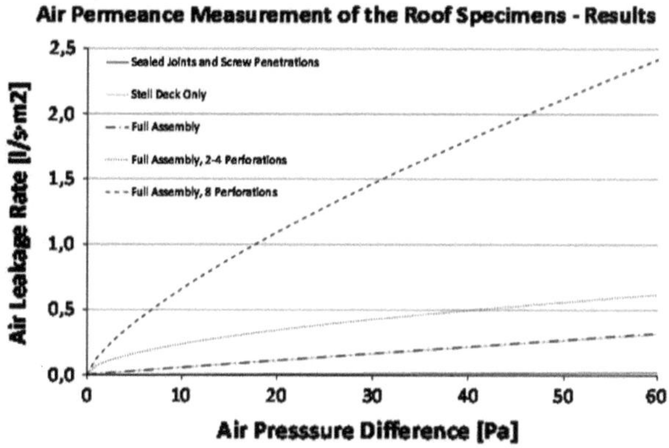


**Fig. 9.6** Increase in air permeance at increasing air pressure differences between the *inner* and *outer* surfaces of the roof assembly. Even small perforations of the steel deck affect the air permeance significantly

insulation boards mounted on top. This assembly was later compared with the full assembly with both steel deck and insulation boards, though without the thermoplastic membrane. The results from these two assemblies indicate that the airtightness of the steel deck is conclusive (i.e., the airtightness of the insulation boards is much less than that of the steel deck). Further, the steel deck was perforated with varying numbers of 3/16-inch drilled holes. The effects of a perforated steel deck are presented in Table 9.1, indicating the importance of intact steel sheeting.

The results from measuring the airtightness of the metal roof construction indicate that even small perforations of the steel increase the leakage rate significantly. A well-performed overlap of the steel deck sheets, screwed tight and without further sealing, has relatively high airtightness, however. Noteworthy is that the tested assemblies do not include any installation or structural penetrations that need to be sealed. The IECC-2012 Standard for Commercial Energy Efficiency declares the importance of sealing penetrations (ICC 2011). Further, the IECC Standard provides three different approaches to determining the air barrier of different materials, construction assemblies, or the complete building, where the latter shall not exceed 2.0 l/s $m^2$ at a pressure difference of 75 Pa. Assuming an n-value of 0.6, this regulation equals a maximum Q50-value of 1.57 l/s $m^2$. Naturally, the air leakage regulation includes all parts of the building envelope and not the roof solely. However, steel decks have been proven leaky, so sealing the joints of the steel sheets and ensuring a non-perforated steel deck is important for overall airtightness (Walsh 2007). A literature study prepared for the California Energy Commission in 2006 presented an average Q50-value of 4.0 l/s $m^2$ based on air leakage tests in 267 commercial buildings (Gadgil et al. 2006).

Another aspect of analyzing the results from the measurements of this study is their expected relationship to constructed roofs. The airtightness of the test specimens and the field measurements have been shown to deviate. Test results of a specimen representing a lightweight compact metal roof without a vapor retarder, like the roof construction presented in Fig. 9.1, resulted in a Q50-value of about 0.75 l/s $m^2$, though field measurements varied between 2 and 6 l/s $m^2$ (Hens et al. 2003).

In the simulation model, the air layer is assumed to have an average thickness of 3 cm, representing the fluttering-induced air cavity.

Further, material properties from the hygrothermal model database are utilized, except for the metal trapezoidal construction where a vapor permeance of 1.0 (perm) is applied. The permeance represents a metal deck including air leakages from penetrations such as screws and is in accordance with earlier investigations done at Oak Ridge National Laboratory (Kyle and Desjarlais 1994). The initial moisture content of the materials in the simulation model are in accordance with EMC80 (ASHRAE 2011).

## 9.4 Parameters

The results of this study are based on multiple iterations of the numerical simulation of the roof construction with varying input parameters. The varying parameters are presented briefly in Sect. 9.2 and are discussed in detail below.

### *9.4.1 Climate*

Four different U.S. climates are used, representing climate zones 4–7:

- Climate Zone 4—Baltimore, Maryland
- Climate Zone 5—Chicago, Illinois
- Climate Zone 6—Minneapolis, Minnesota
- Climate Zone 7—Fargo, North Dakota

The chosen climate of each city represents the 10th percentile coldest climate and is presented with hourly values of temperature, relative humidity, wind speed, solar intensity, etc. These climates are applied to serve as design reference years for the estimation of hygrothermal performance in buildings (Sanders 1996).

### *9.4.2 Surface Solar Reflectance*

The solar absorptivity of a surface defines the ratio of solar radiation absorbed by a surface and varies between 0 and 1. The solar absorptivity used in the simulations of this study is set to either 0.3 or 0.85, which is representative of a white or a dark surface respectively. These values assume that a white roof reflects 70 % of the sunlight and a dark surface reflects only 15 %. The variation in solar absorptivity allows the hygrothermal performance of a cool roof color to be compared to that of a traditional darker roof color. The disparity between the chosen solar absorptivity for the light and dark surface in this study is less extreme than in the literature (Bludau et al. 2009), where 0.2 for a white and 0.9 for a black surface are assumed.

### *9.4.3 Indoor Moisture Supply*

Four different variations of indoor relative humidity are used in the simulations. Due to the lack of specific design values for commercial buildings, equivalent rates for residential buildings are applied. The four different variations of indoor moisture supply are presented here:

- EN-15026, where the indoor relative humidity is assumed to vary with the outdoor temperature and is presented in two different classes, normal and high moisture load (Standardization 2007). The high moisture load of EN-15026 is equivalent to the simplified default indoor design humidity method of ASHRAE 160 (ASHRAE 2011). This study uses both the normal and the high indoor moisture load.
- ASHRAE 160 intermediate method, where the moisture load depends on the specified number of bedrooms and ventilation *ACH*. Two different scenarios are chosen for this study: two bedrooms with $ACH = 0.6$ and six bedrooms with $ACH = 0.2$, which supposedly are representative of a low and a high indoor moisture supply.

### *9.4.4 Air Intrusion*

Typical air intrusion rates are defined in Sect. 9.3.2. The chosen rates are based on four different leakage rates of the roof assembly, presented below. The air intrusion rates used in the simulations of the roof are as follows.

- $Q50 = 0.27$ (l/s $m^2$) A perfectly assembled roof construction with regard to both material properties and workmanship, according to measurements of the air leakage in the roof construction below the thermoplastic membrane in Table 9.1.
- $Q50 = 0.56$ (l/s $m^2$) An assumed satisfactory assembly of roof construction, though with minor perforations in the steel deck; it is based on measuring the average leakage rate with a varying number of 3/16-inch holes (see Table 9.1).
- $Q50 = 1.0$ (l/s $m^2$) Semi-leaky roof construction, arbitrarily chosen.
- $Q50 = 2.0$ (l/s $m^2$) Leaky roof construction, based on both measurements presented in Table 9.1 and air leakage tests made on commercial buildings with similar roof assemblies (Gadgil et al. 2006; Hens et al. 2003).

The pertinent air leakage coefficient, C, and pressure exponent, n, values are taken from either Table 9.1 or are calculated by assuming $n = 0.65$, which in lieu of provided values usually is a good assumption (Gadgil et al. 2006).

## 9.5 Result

A total of 128 different scenarios of the mechanically attached roof system are simulated, with varying input parameters as defined in Sect. 9.4. Each scenario is numerically simulated for a complete year.

### 9.5.1 *Risk of Condensation*

The accumulation of moisture in the air layer between the thermoplastic membrane and the insulation board is evaluated since it is closely related to the amount of condensed water in the roof construction. The moisture content of the air layer is converted into a condensate layer thickness, dl, in which the moisture content is assumed to be distributed over the complete surface area. The conversion is made to facilitate hygrothermal analysis of the roof construction and comparisons between the different simulated scenarios. The variations of dl for each simulation are presented in Fig. 9.9. Typically, dl increases during the heating season, when the difference between the indoor and outdoor moisture content is the greatest.

The variations of dl, illustrated in Fig. 9.9, give a range of different simulated roofs, without specifying the chosen input parameters of the simulations. The maximum values of dl for the simulations are given in Figs. 9.7 and 9.8, as a function of the chosen indoor moisture supply. A curve, with either a black or a white surface, is presented for each moisture supply with varying climate zone and air intrusion rate.

A critical dl is commonly taken as 0.5 mm to avoid dripping (DIN 2001; Hens et al. 2003), hence this value is considered an upper maximum for a safe and reliable roof construction. Additionally, a dl between 0.5 and 1.0 is considered risky, and values beyond are rated as failures in terms of the risk for condensation. The 1.0 threshold is also recommended in German Standard (DIN 2001); it is stated as an upper limit to avoid gravitational flow, but in presence of hygroscopic materials only. However, the standard is mainly used for the surfaces of walls with a sloped roof; consequently, water drips more easily compared to a flat roof. Hence the 1.0 threshold can be assumed conservative and thus applicable for this study.

The results of the risk evaluation for the 128 simulations are given in Table 9.2.

In eight of the 128 roof scenarios, additional simulations were required for a risk evaluation. These eight roof scenarios accumulated moisture, indicating that the annual variation in moisture content will escalate. To confirm this assumption, the eight scenarios were simulated for 5 years, under the same condition but with an additional climate, instead representing the 10th percentile warmest climate. Two of the eight re-simulated roof scenarios reached risky levels and their status has been implemented into Table 9.2.

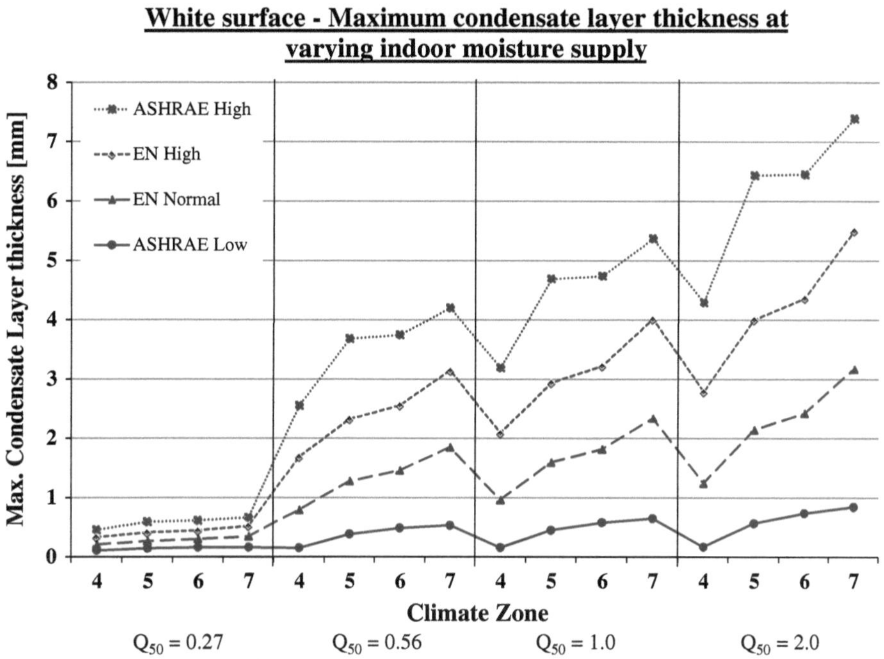


**Fig. 9.7** Maximum condensate layer *thickness* for each simulated roof construction with a *light* membrane surface. The *curves* represent the chosen indoor moisture supply with varying climate and air intrusion rate

## *9.5.2 Energy Performance*

The main reason behind a cool roof retrofit is to decrease the cooling energy demand. However, as a result of the retrofit, the heating demand can increase, since both aspects should be considered in an energy efficiency analysis. In addition, the generation of indoor heat and the thermostat set-point temperatures play significant roles in an analysis of potential energy savings. Figure 9.10 presents the relative energy savings if changing the solar absorptivity of the studied roof construction from 0.85 to 0.30 i.e. from a dark to a light surface color. The potential energy savings are presented as an annual average value and are separated into heating, cooling or the combined heating and cooling energy demand for the U.S. climate zones 4–7. Further, the impact due to the ranges of thermostat temperatures is evaluated, in which two different ranges of set-point temperatures are studied; a small span of 21.1/23.3 °C (70/74 °F) and a wider span of 20/25.6 °C (68/78 °F). According to Fig. 9.10, the impact of the thermostat temperatures is most significant on the energy cooling demand, though proven rather insignificant in heating or when considering both heating and cooling. The most important conclusion of the results in Fig. 9.10 is that the energy saving potential is negative for the U.S. climate zones 6 and 7; however, under conditions with no assumed indoor heat supply.

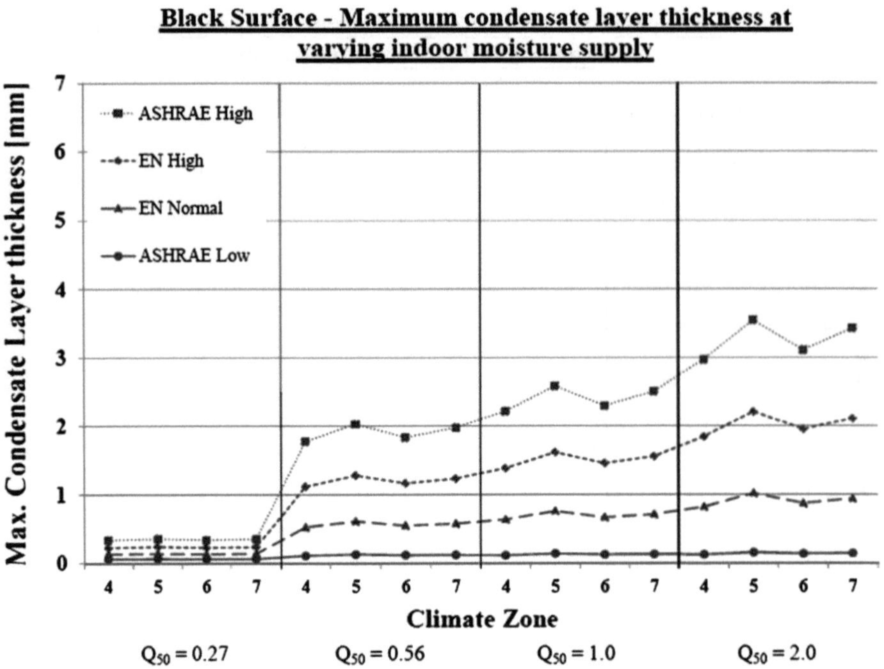


**Fig. 9.8** Maximum condensate layer *thickness* for each simulated roof construction with a *dark* membrane surface. The *curves* represent the chosen indoor moisture supply with varying climate and air intrusion rate

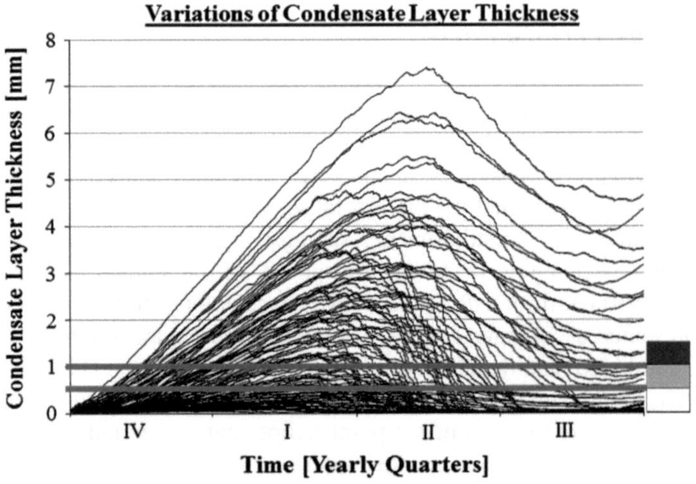


**Fig. 9.9** Condensate layer *thickness*, dl, for the 128 simulated roofs with a mechanically attached *outer* membrane. Typically, the *thickness* increases during the heating season. In accordance with Table 9.2, a considered risky or failing roof assembly, in terms of condensation, is indicated with the color of *grey* or *black*. If the annual variation of dl, at any time, reaches above 0.5 mm, it is referred as risky. The simulated roof assembly is considered a failure if dl reaches 1.0 mm or more

**Table 9.2** Results from the 128 simulated scenarios, indicating the reliability of the roof construction at given conditions

| Climate zone 4 | | | | | | | | |
|---|---|---|---|---|---|---|---|---|
| **Indoor moisture supply** | $Q_{50} = 0{,}27$ | | $Q_{50} = 0{,}56$ | | $Q_{50} = 1{,}0$ | | $Q_{50} = 2{,}0$ | |
| ASHRAE - Low | B | W | B | W | B | W | B | W |
| EN - Normal | B | W | B | W | B | W | B | W |
| EN - High | B | W | B | W | B | W | B | W |
| ASHRAE - High | B | W | B | W | B | W | B | W |
| **Climate zone 5** | | | | | | | | |
| **Indoor moisture supply** | $Q_{50} = 0{,}27$ | | $Q_{50} = 0{,}56$ | | $Q_{50} = 1{,}0$ | | $Q_{50} = 2{,}0$ | |
| ASHRAE - Low | B | W | B | W | B | W | B | W |
| EN - Normal | B | W | B | W | B | W | B | W |
| EN - High | B | W | B | W | B | W | B | W |
| ASHRAE - High | B | W | B | W | B | W | B | W |
| **Climate zone 6** | | | | | | | | |
| **Indoor moisture supply** | $Q_{50} = 0{,}27$ | | $Q_{50} = 0{,}56$ | | $Q_{50} = 1{,}0$ | | $Q_{50} = 2{,}0$ | |
| ASHRAE - Low | B | W | B | W | B | W | B | W |
| EN - Normal | B | W | B | W | B | W | B | W |
| EN - High | B | W | B | W | B | W | B | W |
| ASHRAE - High | B | W | B | W | B | W | B | W |
| **Climate zone 7** | | | | | | | | |
| **Indoor moisture supply** | $Q_{50} = 0{,}27$ | | $Q_{50} = 0{,}56$ | | $Q_{50} = 1{,}0$ | | $Q_{50} = 2{,}0$ | |
| ASHRAE - Low | B | W | B | W | B | W | B | W |
| EN - Normal | B | W | B | W | B | W | B | W |
| EN - High | B | W | B | W | B | W | B | W |
| ASHRAE - High | B | W | B | W | B | W | B | W |

Table cells with no background color indicate a safe roof construction, gray indicates risky construction, and black indicates an expected failure with respect to condensation. B stands for a black roof surface, and W stands for white

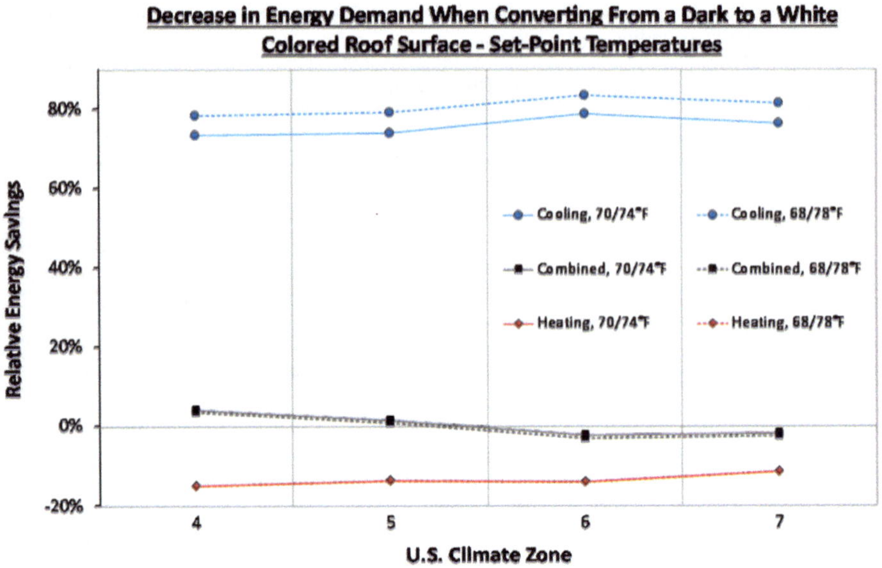


**Fig. 9.10** The decrease in energy demand when converting from a *dark* to *light* colored roof surface. The variations are separated into cooling, heating and combined energy heating and cooling demands. The values represent the U.S. climate zones 4–7 and with two different ranges of thermostat set-point temperatures. A negative value of the relative energy savings equals an increase in energy demand. The reason why the *curve* for the combined heating and cooling demand is far closer to the *curve* for heating demand than for cooling, is that the energy demand for heating is higher than for cooling in the studied U.S. climate zones

Apparently, the impact of the indoor heat supply is more significant than the thermostat temperatures. In Fig. 9.11, three different scenarios of indoor heat generations are considered; no heat supply, 2 °C (4 °F) and 5 °C (10 °F) heat supply. According to the presented results, a cool roof retrofit in climate zone 6 and 7 are still questionable if assuming an indoor heat supply from appliances, humans and solar radiation of 2 °C (4 °F). In cases with an assumed high indoor heat supply, the potential energy savings is positive for all simulated U.S. climate zones (Fig. 9.11).

## 9.6 Discussion and Conclusion

This study focuses on the risk of condensation in cool roof construction with a mechanically attached outer membrane. The risk evaluation is based on 128 simulated scenarios of plausible roof conditions with varying indoor and outdoor climates. Four input parameters are chosen to vary: the outdoor climate, the solar surface properties, the indoor moisture supply, and the indoor air intrusion rate below the surface membrane.

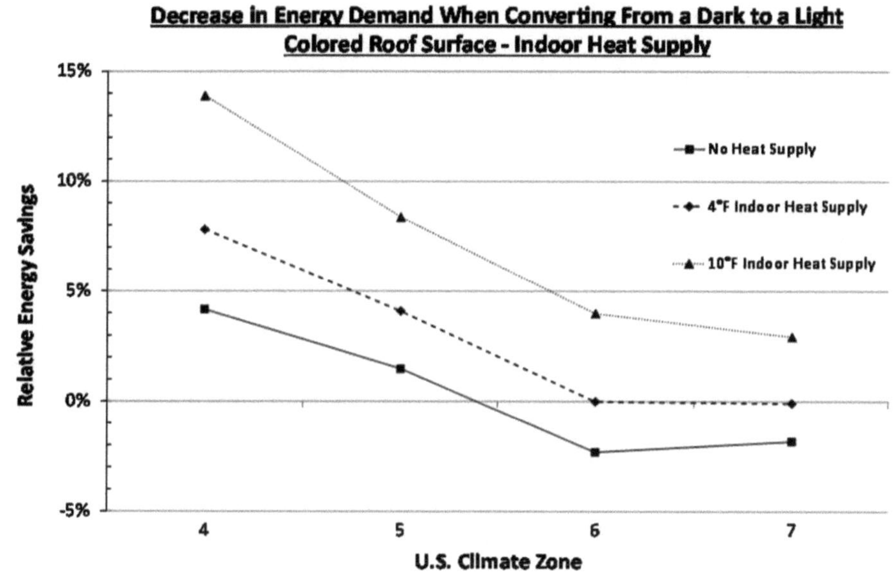


**Fig. 9.11** The decrease in the combined heating and cooling energy demand when changing the solar absorptivity of the surface membrane from 0.85 to 0.30. The variations are presented for the U.S. climate zones 4–7 and with varying indoor heat supply. A negative value of the relative energy savings equals an increase in energy demand

The results emphasize the importance of solar reflectance at the roof surface. Comparing the maximum condensate layer thickness, $d_l$, in Figs. 9.7 and 9.8 reveals that the amount of accumulated moisture is almost doubled in a cool roof construction compared to a traditional black roof.

The indoor moisture supply is very much related to the expected hygrothermal performance of the roof. Referring to maximum $d_l$ in Figs. 9.7 and 9.8, a low and a high moisture supply can cause as much as a 10× difference in condensate layer thickness. Another indicator is presented in Table 9.2, which reveals that only about 18 % of the simulated roof construction is considered risky when the indoor moisture supply is at a low level. (Typically, a low indoor moisture supply is either attained by a low moisture production rate or by a high ventilation air exchange rate.) A distinction between a white and a black surface, for a low moisture supply, shows that only a white surface can be risky with a high air intrusion rate. The opposite is valid for a white surface with a high indoor moisture supply, where only about 16 % of the roof constructions are considered safe. Consequently, the considered safe roof constructions, at high indoor moisture supplies, have black surfaces only.

The different air intrusion rates show similar behavior in comparison with indoor moisture supply. An increased rate increases the maximum $d_l$, which also can be seen by the increased slope of the curves in Figs. 9.7 and 9.8 and by comparing the vertical sections of Table 9.2.

The fourth varying parameter, the climate, was proven to also have an influence on the amount of accumulated moisture. A distinguished slope between each climate section in Figs. 9.7 and 9.8 confirms an increase in the maximum $d_l$. The influence of the climate is also obvious by comparing each climate segments of Table 9.2.

In conclusion, all the varying parameters of this study are highly influential on the hygrothermal cool roof performance. A low indoor moisture supply or a low air intrusion rate ensures a low risk of intermediate condensation. A safe upper limit of air leakage at 50 Pa, $Q_{50}$, is stated as 0.17 l/s m$^2$ for metal roofs (Hens et al. 2003). This limit seems consistent with the low risks of the simulated lower air intrusion rate of 0.27 l/s m$^2$, at least for the black roofs of this study. Further, a cool roof will accumulate approximately twice as much moisture below the surface membrane as a black surface. It is assumed that replacing a black surface with a cool membrane on an existing mechanically attached roof system could result in intermediate condensation. The impact of the climate on the cool roof hygrothermal performance is verified and a conclusion can be made based on the results presented in Figs. 9.7, 9.8 and Table 9.2; apparently, a colder climate increases the risk of condensation.

An analysis of the energy efficiency potential reveals that the cooling energy demand is significantly lowered by a cool roof retrofit, though simultaneously, the heating demand is increased. Therefore, the combined heating and cooling demand ought to be included in an energy efficiency analysis. Apparently, a cool roof retrofit in U.S. climate zone 6 and 7 can actually have a negative effect on the energy demand if no, or very low, indoor heat generation is expected.

The mechanical resistance of the roof membrane has not been taken into account in this study. Likely, this means that, at some lower limit of wind-induced pressure, the uplifting force is lower than the weight and flexible resistance of the membrane, thus preventing any air intrusion. Therefore, a complete depressurization analysis of a mechanically attached roof system is needed to fully analyze a cool roof assembly at realistic and fluctuating wind loads.

It is of great concern to emphasize that a single ply roof, including an interior vapor retarder, is not necessarily equivalent with an airtight construction. Either insufficiently sealed overlaps, perforations or penetrations of the vapor retarder, may cause high air intrusion rates.

Finally, the following practical conclusions can be stated:

- If a very low indoor moisture supply is assumed, no moisture problem is expected, except for white surfaces combined with high air intrusion rates.
- For black roofs, the joints of the steel deck do not necessarily need to be sealed to be considered safe, though penetrations and perforations must.
- The previous statement is valid for white roofs, only with a low or normal indoor moisture supply.
- For all other roof assemblies with varying indoor and outdoor climates, an interior air barrier is recommended.

## References

ASCE (2010) Minimum design loads for buildings and other structures ASCE/SEI 7-10. American Society of Civil Engineers

ASHRAE (2011) ANSI/ASHRAE standard 160-2009 criteria for moisture-control design analysis in buildings. American Society of Heating, Refrigerating and Air-conditioning Engineers, Inc., Atlanta, GA

ASTM (2011a) D7586/D7586M-11 standard test method for quantification of air intrusion in low-sloped mechanically attached membrane roof assemblies. ASTM International

ASTM (2011b) E2178-11 standard test method for air permeance of building materials. ASTM International

Baskaran BA, Molleti S (2010) How much air is too much? The National Research Council of Canada studies roof system air intrusion. NRC Publications Archive

Bludau C, Zirkelbach D, Kuenzel HM (2009) Condensation problems in cool roofs. Interface. J RCI XXVII(7):11–16

Davis FK, Newstein H (1968) The variation of gust factors with mean wind speed and with height. J Appl Meteorol 7(3):372–378

DIN, D. I. f. N (2001) DIN 4108-3, Teil 3: Klimabedingter Feuchteschutz, Anforderungen, Berechnungsverfahren und Hinweise für Planung und Ausführung *Wärmeschutz und Energie-Einsparung in Gebäuden*. DIN Deutsches Institut für Normung

Ennis M, Kehrer M (2011) The effects of roof membrane color on moisture accumulation in low-slope commercial roof systems. Paper presented at the proceedings of NRCA international roofing symposium 2011, Washington, MD

Gadgil A, Price PN, Shehabi A, Chan R (2006) Indoor-outdoor air leakage of apartments and commercial buildings: public interest energy research (PIER) program. California Energy Commission

Hagentoft C-E (2001) Introduction to building physics. Studentlitteratur, Lund

Harper BA, Kepert JD, Ginger JD (2010) Guidelines for converting between various wind averaging periods in tropical cyclone conditions. World Meteorological Organization (WMO)

Hens H, Zheng R, Janssens H (2003) Does performance based design impacts traditional solutions? Metal roofs as an example. Paper presented at the proceedings of the 2nd international conference on building physics, Antwerpen, Belgium

ICC (2011) 2012 International Energy Conservation Code®: International Code Council, Inc

Kehrer M, Schmidt T (2008) Radiation effects on exterior surfaces. Paper presented at the proceedings of Nordic symposium on building physics 2008, Copenhagen

Kyle DM, Desjarlais AO (1994) Assessment of technologies for constructing self-drying low-slope roofs. Martin Marietta Energy Systems, U.S. Department of Energy, Oak Ridge, TN

Künzel HM (1995) Simultaneous heat and moisture transport in building components. One- and two-dimensional calculation using simple parameters. Dissertation, University Stuttgart, IRB Verlag. Retrieved from www.WUFI.com

Künzel HM, Zirkelbach D, Scfafaczek B (2011) Vapour control design in wooden structures including moisture sources due air exfiltration. Paper presented at the 9th Nordic symposium on building physics—NSB 2011, Tampere, Finland

Molleti S, Baskaran B, Kalinger P, Beaulieu P (2011) Air intrusion and its effect on moisture transport in mechanically attached roof systems. Paper presented at the proceedings of the 2011 international roofing symposium

Sanders C (1996) Annex 24 heat, air and moisture transfer in insulated envelope parts, environmental conditions. Final report, volume 2, ECBCS Annex Publications

Standardization ECf (2007) EN-15026 hygrothermal performance of building components and building elements—assessment of moisture transfer by numerical simulation: ON Österreichisches Normungsinstitut

Walsh O (2007) The essential guide to Part L of the building (amendment) regulations 2007. Navitus Energy Consultants, Dublin, Ireland

MIX
Papier aus verantwortungsvollen Quellen
Paper from responsible sources
FSC® C105338
FSC
www.fsc.org

If you have any concerns about our products,
you can contact us on
**ProductSafety@springernature.com**

In case Publisher is established outside the EU,
the EU authorized representative is:
**Springer Nature Customer Service Center GmbH**
**Europaplatz 3, 69115 Heidelberg, Germany**

Printed by Libri Plureos GmbH
in Hamburg, Germany